Stampedia Philatelic Journal 2020

無料世界切手カタログ・スタンペディア株式会社

Publisher：Stampedia,inc.

President & Editor in Chief：YOSHIDA Takashi

Date of Issue：10th Nov. 2020

Number of Issue：600

Price：1,000 Yen in Japan, 10Euro or $12 overseas

目次
Table of Contents

序文

Introduction

当雑誌の発行にご賛同くださいます広告主の皆様、そしてご購読くださる皆様に御礼申し上げます。

オンライン世界切手カタログ・スタンペディアを運営する私が「Philatelic Journal」という書籍を2011年春に発行した意図は、日本で十年ぶりに開催される国際展 PHILANIPPON2011 を盛り上げたい一心からでした。国際展出品経験のある日本のフィラテリストにご寄稿をいただき、日本のフィラテリーの水準を PHILANIPPON2011 の直前に世界に発信することがかない、その目的を多少ですがかなえることができました。

当初は一冊限りの発行と考えておりましたが、ご購読いただいた皆様から葉書やメールで多くの励ましを頂くと共に、FIP 国際切手展・文献部門で雑誌としては異例の大金銀賞（2017 年からは金賞）という高い評価を頂く事ができ、年一度の刊行を続けてみようと考えるに至りました。

今年は 5 名のフィラテリストの方々からすばらしい記事をご寄稿いただきました。感謝申し上げます。2012 年より日本人に限らず幅広い記事の依頼をしており、基本的にどの記事も当誌の為に書き下ろして頂いた物で読み応えがあります。どうぞお楽しみください。

We appreciate all the advertisement supporter as well as every philatelist who generously purchased STAMPEDIA PHILATELIC JOURNAL. I'm a Japanese philatelist as well as a founder of "Stampedia" : the Whole Earth Stamp Catalogue on the internet".

The reason I published "STAMPEDIA PHILATELIC JOURNAL" in April 2011 is because I'd like to appeal to philatelists all over the world the attractiveness of the PHILANIPPON2011, the first Japanese FIP show in the decade. Thanks to many top Japanese philatelists, I achieved the objective by appealing the high level of the philately in Japan.

I thought to issue such a book only once in 2011, however I changed my mind after receiving many emails and postcards from the readers encouraging me, and as a result I came to an idea to publish it every once a year.

We have 5 contributors this year, and I dare say every article is super content. Please enjoy.

Tokyo, November 2020

YOSHIDA Takashi

世界切手展 受賞履歴
Awards at the F.I.P. World Stamp Show

PRAGA 2018 GOLD
WORLD STAMP SHOW ISRAEL 2018 GOLD
BRASILIA 2017 GOLD
WORLD STAMP SHOW NEW YORK 2016 LV + SP

It's our pleasure for our philatelic magazine to be awarded many times as GOLD award, which is the highest medals given to the philatelic magazine / periodical section in the literature exhibits by now. We're also going to do our best to promote philately in the future.

当誌が雑誌部門に対する評価として最も高い金賞に国際展で何度も輝く事ができたのは望外の喜びです。これからも郵趣振興の為に頑張ります。

COLLECTION

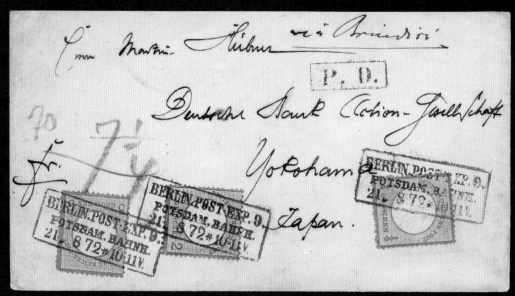

Mail to Japan – German Empire 1872, small shield mixed franking on cover from
Berlin to the first overseas branch of Deutsche Bank in Yokohama (opened 1872)

21 November 2020

German States · Shield Issues – The ERIVAN Collection– 4th Sale

www.heinrich-koehler.de

Order the auction catalogue now!

+49 - (0)611 - 34 14 9-0 · info@heinrich-koehler.de

HEINRICH KÖHLER

Germany's Oldest Stamp Auction House

FOUNDED IN 1919

CORINPHILA - TRADITION

AND EXPERIENCE

IN CLASSIC PHILATELY

OUTSTANDING RESULTS FROM RECENT CORINPHILA SALES 2007 - 2020 *

CHF 788,700   **CHINA 1897**, *1 dollar mint block of 15 (October 2008)*

CHF 720,000   **BRAZIL 1843**, *60 reis, the unique mint sheet (June 2013)*

CHF 605,000   **CHINA 1897**, *1 dollar ‚Small Dollar' mint (June 2018)*

CHF 573,600   **SWITZERLAND 1850**, *The ‚Winterthur' block of 8 on cover (June 2009)*

CHF 523,600   **CHINA 1897**, *5 dollar mint pair with inverted overprint (December 2007)*

CHF 406,300   **ZURICH 1843**, *'4' mint strip of 5 (June 2009)*

CHF 384,000   **BASLE 1845**, *the ‚Renan Cover' (June 2017)*

CHF 360,000   **BRAZIL 1843**, *30 reis interpane block of four (March 2013)*

CHF 334,600   **CHINA 1897**, *2 cents, mint sheet of 100 (October 2008)*

CHF 324,000   **GENEVA 1846**, *Large Eagle block of 20 (April 2012)*

CHF 314,600   **WESTERN AUSTRALIA 1854**, *4d., 'Inverted Swan' (June 2018)*

CHF 262,900   **AUSTRIA 1850**, *'Yellow Mercury' mint pair (October 2008)*

CHF 259,600   **SWITZERLAND 1850**, *'Waadt 5' and Rayon II (2) on cover (Febr. 2007)*

CHF 239,000   **CANADA 1851**, *12 Pence mint (March 2010)*

CHINA 1897,
"Small Dollar – The King of Chinese Philately"

CHF 605,000.- *
June 2018

Western Australia, "Inverted Swan"
The most valuable stamp of Australia

CHF 314,600.- *
June 2018

PLUS ANOTHER 40 REALISATIONS *
BETWEEN CHF 100,000 AND 380,000 !

BAKER TILLY OBT AG
All realisations in 2007 - 2019 over
CHF 100,000 hammer prices
confirmed by
Certified Swiss Accountant !
Full Accountant's report online on
www.corinphila.ch

* Hammer Prices incl. Buyer's Premium (excl. tax)
Exchange rate correct in October 2020: CHF 1 = US$ 1.10

CORINPHILA AUKTIONEN AG
VIESENSTR 8 · 8032 ZURICH · SWITZERLAND
Phone +41-44-3899191
www.corinphila.ch

CORINPHILA VEILINGEN BV
AMSTELVEEN · NETHERLANDS
Phone +31-20-6249740 · www.corinphila.nl

CORINPHILA AUCTIONS

As the oldest stamp auction house in Switzerland, situated in the international financial centre of Zürich, we at Corinphila really know the market.

The most specialised philatelic knowledge, fastidious presentation and an international customer base with strong purchasing power guarantee the highest prices.

We are quite willing to discuss larger holdings in your own home.

YINGKOW
1932年8月16日 →
I. J. P. O.

CHANGCHUN
1932年8月17日 →
I. J. P. O.

CHANGCHUN
1932年8月18日 → HARBIN
長春

昭和6年（1931）9月18日、満州事変。
　昭和7年（1932）3月1日、「満州国」の建国。
「満州国」政府は、建国時の中華郵政の郵便料金をそのまま適用した。
　昭和7年（1932）7月26日、「満州国」切手を発行。旧中国切手は、8月25日まで有効（但し9月7日まで黙認）。「満州国」が中華郵政を引き継ぐと、「満州国」内日本局から「満州国」に差出される郵便は「満州国」切手への貼替えを要した。このため上記の例は「満州国」切手ではなく、長春局で旧中国切手に貼替えられている。

第一種：4分×2 ＋ 書留：10銭 ＝ 18銭

↓貼替え

信　函：4分×2 ＋ 掛號： 6分 ＝ 14分

裏面コピー ×50%

ブース出店のご案内

第58回 世界の貨幣・切手・テレホンカードまつり：（株）新橋スタンプ商会主催
2020年12月4日（金）〜12月6日（日）東京交通会館12階ダイヤモンドホール

第34回 JSDA 切手まつり：日本郵便切手商協同組合主催
2021年3月19日（金）〜21日（日）東京交通会館12階ダイヤモンドホール

株式会社 ユキオ・スタンプ。

店舗：〒460-0003 名古屋市中区錦三丁目16番10号先
栄森の地下街　北二番街
TEL&FAX (052) 211-7587

営業時間：11時〜18時
通信先：〒486-8691 春日井郵便局私書箱6号

富士鹿切手の製造面

Japan Definitive issue 1922-1937, and its Production
Landscape Stamps for UPU Surface Rates
吉田 敬 YOSHIDA Takashi

富士鹿切手の概要

　風景デザインの切手は、英領植民地を中心として、２０世紀を迎える頃から世界的に流行しはじめました。この潮流をうけて発行された日本初の風景意匠切手が富士鹿切手です。

　そもそもこの切手は、1920年の第7回UPU総会(マドリッド)における加盟各国の外国郵便料金の値上げについての決定をうけて発行された切手です。

　その決定内容は、

(1) 今までの郵便料金を下限とし、その二倍の金額までの値上げを実施する。
(2) 通貨変動に従う料金変更を可能とする。

　というもので、日本の外国郵便料金(船便)は、10銭、4銭、2銭(書状、葉書、印刷物の順)から20銭、8銭、4銭に、1922年1月1日より改められることになりました。

　ところが、既に発行されていた20銭、8銭、4銭切手(fig.1)の刷色はUPU推奨の刷色ではなかったため、新切手の発行が必要とされ、その意思決定の過程で、意匠は既存意匠の流用でなく、新意匠の採用が決定されたものです。

　第7回UPU総会(1920)の決定を受けて、外国郵便料金用の切手を、それ以外の切手と別意匠で発行した事例としては、フランスのパストゥールシリーズが有名です。(fig.2)富士鹿切手は、このパストゥール切手と同等の重要性のある日本切手と言えます。

Overview of the 1922-1937 issue

Landscape design stamps became popular mainly in the British colonies from the beginning of the 20th century. The 1922-1937 issue was the first Japanese landscape stamp following this tide.

It was issued according to the foreign postage rise decided at the 7th UPU Congress (Madrid) in 1920, in which they agreed:

(1) each country of the Union raise the existing foreign postage with the limit of the double.

(2) each country of the Union may modify the postage following the fluctuations of currency.

In Japanese postal service (surface mail), the postage changed on January 1, 1922, from 10 Sen, 4 Sen, and 2 Sen (letter, postcard, and printed matter, in this order) to 20 Sen, 8 Sen, and 4 Sen.

However, the printing color of already issued stamps of 20 Sen, 8 Sen, and 4 Sen (fig.1) was not the UPU's recommended color. So, they needed to issue new stamps and applied the new design in the decision-making process.

There were other cases of changing the international stamp design according to the 7th UPU Congress agreement in 1920. The famous one was the French Pasteur issue (fig.2). The 1922-1937 issue is such a valuable Japanese stamp as the Pasteur issue.

fig.1 当時の 4 銭、8 銭、20 銭
4 Sen, 8 Sen, and 20 Sen stamps in 1920

fig.2 1923 仏通常切手パストゥールシリーズ
1923 France Pastour issue

富士鹿切手の一覧

外国郵便料金用の切手として発行が開始された富士鹿切手は、国内での使用を前提として発行された普通の切手に比べて馴染みが薄く、その結果としてカタログ価格が高くなりがちです。また発行数が少ないため、製造面のバラエティも少ないと過去には考えられてきました。

両者の悪循環で、シートや大型マルチプルを用いた製造面研究がされることも少なく、日本の通常切手の中では、伝統郵趣作品を作ることが難しいシリーズと考えられてきました。

しかし市場に残存していたわずかなシートや大型マルチプルの大半を集めて調査してみると、これまでに発表されていない製造上のトピックスが多く発見されました。

次ページ以降で額面別に解説していきたいと思いますが、その前に、まず普及している日本切手カタログにおいて、富士鹿切手がこれまでどのように分類されてきたかを示します。(tbl.1)

旧版富士鹿切手の銘版違いは、メインナンバーでは分類されていませんので、旧版富士鹿は3種にしか分類されていませんが、改色富士鹿、昭和毛紙、昭和白紙には合計で11のメインナンバーが与えられており、合計で14種類になります。

The 1922-1937 issue Lineup

The 1922-1937 issue, intended for international mail, was less common than other domestic purpose stamps. So, it tends to have a high catalogue price. And since the issued number was not large, it had been thought to have only a few varieties in manufacture.

The vicious spiral of price and the small issued number cause less study in manufacture with sheets or large multiples. And it had been believed a challenging series to make a traditional collection of all the Japanese definitives.

However, by studying sheets and large multiples only surviving a few in the market, I have found many manufacturing topics that had never been reported.

Before explaining the details, I show how the popular Japanese stamp catalogues have classified the 1922-1937 issue by now (tbl.1).

The main numbers don't include the old die's imprint varieties, so the 1922 issue has only three main numbers. And the other issues, such as the Color Change, the Showa Granite Paper, and the Showa White Paper, have eleven main numbers in all. There are fourteen main numbers in total.

シリーズ名 Name of the issue	額面種類 Number of face values	メインナンバー数 Main numuber	DIE	製造時期 Production Period	よく知られた分類法 Well known classification
旧版富士鹿 1922 issue	3	3	I	1921-1929	製造時期により、銘版を4種に分類できる 4 imprint varieties
改色富士鹿 1929 issue	3	6	I, II	1929-1936	旧版と新版の2種に分類できる。 2 DIE varieties
昭和毛紙、昭和白紙 1937 issue	3	5	II	1937-1939	用紙を2種に分類できる。なお、8銭改色の白紙（新版）は、ここに分類されている。 2 kinds of papers, 8 Sen brown white paper is classified into this issue in traditional classification.

tbl.1 普及している日本切手カタログにおける富士鹿切手の分類
Classification in popular stamp catalogue in Japan

一方で、富士鹿切手を額面別にみると tbl2 のようになります。次頁以降では、この表に従い、4銭の製造面を詳解した後、残りの額面を解説してまいります。

tbl.2 is a list of 1922-1937 issue by face value. I start this article by describing the manufacture of 4 Sen stamps, then other face values.

額面 Face Value	旧版富士鹿 1922 issue produced in 1921-1929	改色富士鹿 1929 issue produced in 1929-1936	昭和毛紙、昭和白紙 1937 issue produced in 1937-1939
4 銭 4 Sen	緑 銘版部分で4分類可能 Green, classified into 4 by imprint	橙 印面寸法で2種に分類可能 Orange , 2 DIEs	緑 用紙を2種に分類可能 Green, classified into 2 by paper
8 銭 8 Sen	赤 銘版部分で4分類可能 Red, classified into 4 by imprint	茶 印面寸法で2種に分類可能 Brown, 2 DIEs	茶 白紙のみ Brown, Only white paper
20 銭 20 Sen	青 銘版部分で4分類可能 Blue, classified into 4 by imprint	紫 印面寸法で2種に分類可能 Purple, 2 DIEs	青 用紙を2種に分類可能 Blue, classified into 2 by paper

tbl.2 富士鹿切手を額面で分類した表
Table classifying the stamps by face values

4 銭

原画

　第 7 回 UPU 総会（1920）の後、逓信省は 1922.1.1 からの郵便料金改定を決定し、新しい外国郵便料金（船便）は、20 銭、8 銭、4 銭（書状、葉書、印刷物の順）と決まり、新しい意匠を用いて発行することが決定しました。

　新しい意匠を作成したのは、樋畑正太郎（樋畑雪湖、逓信博物館員）でした。田沢切手の 2 銭緑と 10 銭青は刷色を変更せずに継続して製造・発行される事が決定していた為、外見で一目見て違いのわかる意匠にする事が求められ、1875 年の鳥切手以来の具体的な主題が描かれた意匠とする事が決定しました。

　この原画は郵政博物館に収蔵されていますが、全シリーズを通じて、現存する唯一のアーカイブです。つまり、プライベートにはアーカイブは一点も存在しません。(fig.3)

4 Sen

Original Drawing

After the 7th UPU Congress (1920), the Ministry of Communications decided to revise the postage rate on January 1, 1922. The new international postage rate (surface) was set as 20 Sen, 8 Sen, and 4 Sen (letter, postcard, and printed matter in this order), and a new design was applied.

The new design was created by Shotaro HIBATA (known as Sekko HIBATA, a member of the Communications Museum). Since 2 Sen green and 10 Sen blue of the Tazawa issue were to be manufactured and issued without changing the ink color, the new design required a different looking. Then, they decided to apply a topical design for the first time after the Bird issue in 1875.

The original drawing in fig.3, kept as a collection of the Postal Museum of Japan, is the only remaining archive of all the 1922-1937 issue. Thus, it means there are no archives in private hands.

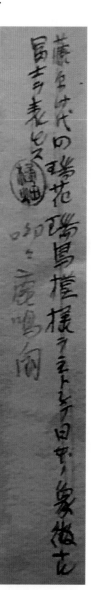

右側の文字の拡大図 , Two vertical writings at the right
.....Mt.Fuji as symbol of Japan...
circled seal HIBATA

fig.3 富士鹿切手の原画　（郵政博物館 収蔵）
Unique original drawing of the issue, kept in the Postal Museum of Japan

原版 DIE I

この原画を元にした原版の製造は、『富士鹿・風景（JPS,1993）』によれば、『銅版に切手の原寸の4倍大に彫刻したもの（これを型版という）をつくり、これを縮刻機にかけて原寸大で銅版の上に彫り込んで行くが、これが原版となる。』とされています。

英語で言えば、『DIE I』と呼べる、この原版からなる印刷版で製造された切手は、1922年の3種に加えて、1929年の3種（改色旧版）も含まれます。これに比べて、1930年以降に導入された切手は『DIE II』からなる印刷版で製造されています。（通称「新版」と呼称される）これをわかりやすく表にすると tbl.3 のようになります。

Original Die, DIE I

The book "1922-1937 issue" (JPS,1993) describes making an original die based on the original drawing. It says, "Engrave (the original drawing) on a copper plate, four-time as large as a stamp size (known as "model plate"), and print it on another copper plate in reduced size, the original stamp size, by the carving machine. The completed plate is an original die."

It is "DIE I." The stamps produced from DIE I are not only three kinds of the 1922 issue but also three stamps of the 1929 issue are included. Stamps were manufactured from "DIE II" (known as "New Die") after 1930. You can see the classification in tbl.3.

原版 DIE	該当する切手 Stamps produced with each DIE		原版の製版方法 DIE producing method	二次原版の要素数 Num. of second DIE
旧版 DIE I	1922年旧版富士鹿シリーズ3種 1929年改色シリーズの旧版3種	1922 issue, 3 kinds 1929 issue DIE I, 3 kinds	エルヘート技法	10 or 25
新版 DIE II	1929年 改色シリーズの新版3種 1937年 昭和毛紙2種 1937年 昭和白紙3種	1929 issue DIE II, 3 kinds 1937 issue granite paper, 2 kinds 1937 issue white paper, 3 kinds	写真製版 Photomechanical process	25

tbl.3 富士鹿切手を原版で分類した表
Table classifying the stamps by DIEs

DIE I と DIE II の区分は切手意匠部分の幅と高さで区別するのが一般的ですが (fig.4)、慣れてくると印面の印象からも判断できるようになります。

Generally, DIE I and DIE II are distinguished by the design area's width and height (fig.4), but you will see them only by the printed face's looking after some experience.

Enlarged View of Difference between DIE I and DIE II

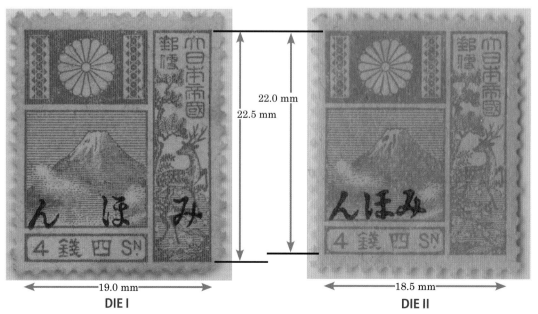

fig.4 旧版 と新版の違い
Differences between DIE I and DIE II

DIE I の二次原版の構成要素数

富士鹿切手は凸版印刷機（平面版）で製造することになっていました。その為、エドアルド・キョソーネが凸版切手の製造にあたり紙幣寮（のちの印刷局）に導入した『電胎法』と呼ばれる手法により、原版を正確に複製して印刷版を作成する手法が使用されたと考えられています。

日本に導入された電胎法では、DIE をシート枚数分転写して印刷版を作るなどという原始的なことはしておらず、中間工程に『第二次・第三次の原版が存在する』事が知られています。

以下に小判切手の第二次原版について記述します。なお、内容は、井上和幸氏が、2012 年に書いた『U・新小判切手におけるブロック収集の難しさと重要性（井上和幸氏、スタンペディアフィラテリックジャーナル 2012)』を参考にしました。

旧小判切手から U 小判切手の初中期までは、原版を元に田型原版を第二次原版として作成し、第二次原版を 20 回（1886 年から 25 回）複写することで、横 10 x 縦 8 の 80 面シート（1886 年から 10 x10 の 100 面シート）の印刷版を作った上で、用紙への印刷を行っていました。

1891 年になると、第二次原版を横に 5 つ繋げた横 10 x 縦 2 の第三次原版が制作され、第三次原版を 20 回複写することで、100 面シートを 4 面掛けした印刷版が使用されたと考えられています。

以後、日本の凸版切手の製造は 100 面シートの 4 面掛け（合計４００枚）で行われたと考えられています。

なお小判切手の時代は 4 面（田型）だった第二次原版は、菊切手の製造以降は、横 5 x 縦 2 の 10 面に変更されており、富士鹿切手も当初は同様でした (fig.5)。

Composition of The Second Die of DIE I

The 1922-1937 issue was to be manufactured by the typography (flat press). So, they must have duplicated the die exactly to make a pane by the "electrolytic etching," the method which Edoardo Chiossone introduced into the Paper Money Bureau (later the Printing Bureau) to manufacture stamps by typography.

The electrolytic etching mothed Japan introduced require to make the 2nd or the 3rd die in the production instead of primitive method of duplicating the DIE for all the subjects in the sheet.

The following is how to make a Second Die of the Koban issue. I referred to the article of Mr. Kazuyuki INOUE, "Difficulty and Importance of Collecting Blocks of U-Koban and New Koban Series" (Stampedia Philatelic Journal, 2012).

From the period of the Old Koban to the beginning and the middle period of the U-Koban, they made a block-of-four die from the original die, namely the Second Die. They did twenty-time duplications of the Second Die (25 times from 1886) to make an 80-subject pane consisting of 10 x 8 (100-subject pane of 10 x10 from 1886) before printing.

In 1891, they put horizontally five Second Dies to make the 3rd die of 10 x 2 and duplicated it twenty times to make a unit of four 100-subject panes.

After then, the Japanese typography stamps are assumed to have been manufactured from a unit of four 100-subject panes (total 400 subjects).

The Second Die, consisting of four subjects (block of four) in the Koban period, changed to a 10-subject die of 5 x 2 from the Chrysanthemum issue production. It continued until the beginning of the 1922 issue (fig.5).

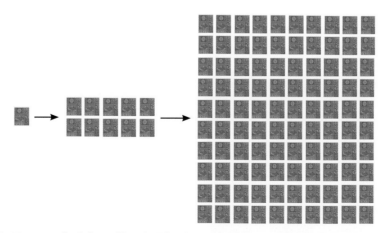

fig.5 左から順に DIE（原版）→ 第二次原版（10 面の場合)→ 印刷版（100 面１枚掛けの場合)
Original die -> The Second Die -> Printing pane

4銭　DIE I 二次原版のキーポジション Type 8　　4 Sen DIE I, The Second Die's Key Position Type 8

　二次原版が、本当に 10 面であるか否かを確認するには、シートや大型マルチプルに当たるのが適切ですが、ここで大きな問題に当たります。初期の富士鹿切手は未使用シートがほとんど残されていないのです。

　次ページの fig.6 は、現存 1 点の初期シートとされる貴重なものです。筆者はこのシートをルーペで注意深く観察した結果、pos.6/20 の 1 0 枚ブロック部分が右上にずれている事に気付き、印刷版は 10 面素版で製造されたことを確信しました。

　しかしながら、そのズレは微妙なものであり、競争切手展で審査員を納得させるには、より明確な何かが必要だとも感じていました。

　そんなことを思いながらシートをぼんやり眺めていたある日、fig.7 の変種が目に飛び込んできました。早速シートの 100 ポジション全てで、この変種の有無をチェックしたところ、fig.6 で赤丸をつけた pos. にのみ変種の存在を確認できました。この結果二次原版の構成数が横 5 枚縦 2 枚の 10 枚であることと、この大きな定常変種はその下段中央であることがわかりました。この定常変種を筆者は「キーポジション Type8」と読んでいます。

　先述の通り、シートや大型マルチプルの残存数が少ない為、DIE I の 3 額面の内、ここまで大きな二次原版の変種が発見できたのは、まだ 4 銭だけですが、今後、他の額面でも見つけていきたいと考えています。

You can confirm whether a Second Die consists of 10-subject dies or not by checking sheets or large multiples; however, you will face a severe problem. Only a few sheets of the 1922 issue and the 1929 issue DIE I remains, especially of its early production. No sheets remain for some of the main numbers.

fig.6 is the only recorded sheet of the first production of 4 Sen of the 1922 issue. I studied it carefully with a magnifying glass and found the block of 10 of the pos.6/20 is slightly shifted to the upper-right. It convinced me that 10-subject second die made this sheet.

However, the shift was slight, and I had to find something more concrete to convince the jury at the competitive exhibition.

One day when I was gazing at the pane thinking like that, the variety in fig.7 suddenly jumped into my eyes. I immediately checked all the 100 positions if they have this variety and saw only the positions circled in red in fig.6 have it. It shows that a Second Die consists of 10 original dies of 5 x 2, and that remarkable constant variety is in the middle of the second line. I named it "Key Position Type 8."

Since there are just a few panes or large multiples, I've found such a remarkable variety only in 4 Sen in all three face values of DIE I. I expect to find one in the other face values.

Key position 8　　　fig.7　　　**Normal position (9)**

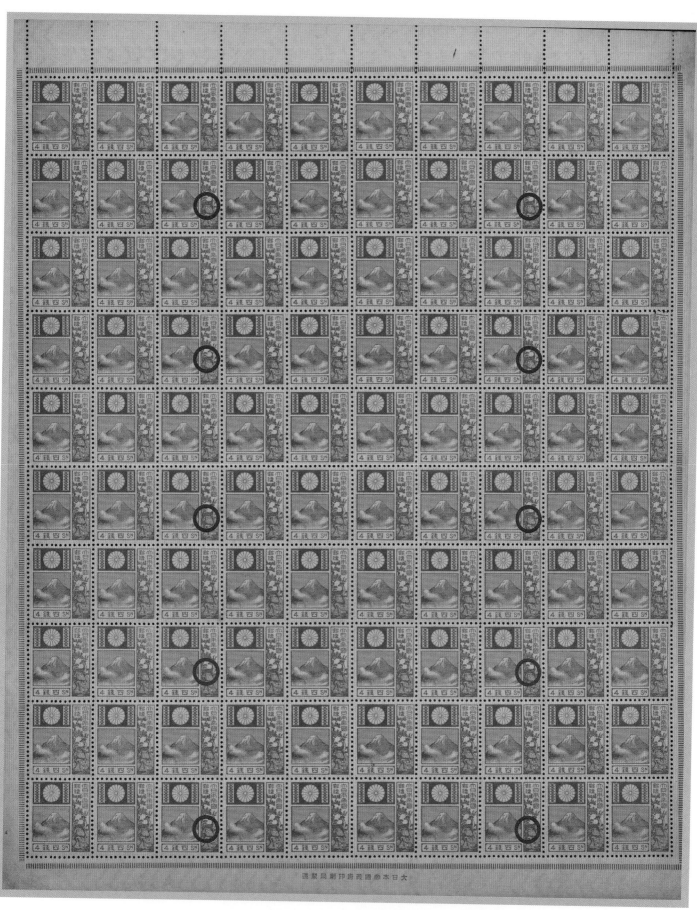

fig.6 1922 年 4 銭切手の初期製造分の未使用シート（現存一点）
10 面素版の Type 8 のポジション全てに定常変種が確認できる。また十字トンボがない為斜めに裁断されていることが分かる。
Only one recorded unused sheet of the 1921-1923 production of 1922 4 Sen
Key position of type 8 of the second DIE can be found in all the expected positions, which are red circled.

DIE I の二次原版構成数の変更

二次原版の構成数が、小判切手の時代は4であったことを先述しましたが、富士鹿切手発行中の1920年代後半に構成数が10から25に変わったと考えられています。

fig.8は、富士鹿切手の大型マルチプルとしては例外的に残存数が多く、10シート近く残っていると考えられる、1922年、旧版富士鹿4銭の第4期の使用済シートです。

このシートで、前ページで触れた「キーポジションType 8」を探すと、Pos.43, 48, 93, 98の4カ所（赤丸のついた部分）にしか出現しないことがわかります。これは、fig.8の製造に使用された二次原版は構成数が横5枚縦5枚の25枚であったことを示しています。

この事から、1922年シリーズは、製造当初は、10面の二次原版で印刷版を作成したが、最終的には25面の二次原版に切り替わったことがわかります。

Change of Composition of the Second Die of DIE I

A Second Die consisted of four original dies in the Koban issue period. While the 1922-1937 issue was circulated, the Second Die consisting of 10 changed to 25 in late 1920's.

fig.8 is a used sheet of the 1922 issue 4 Sen. It's classified to the fourth production and manufactured between 1926-1929. About 10 sheets remains for the 1926-1929 production of the 4 Sen, which are the only exceptions of rare remaining of sheets of the 1922 issue.

In the pane in fig.8, you can find "Key Position Type 8" only in four positions such as Pos.43, 48, 93, and 98, which are red circled. It means that the Second Die used for the pane in fig.8 was consisted of 25 dies of 5 x 5.

This fact shows that the 1922 issue production started with the Second Die of 10 and changed to 25.

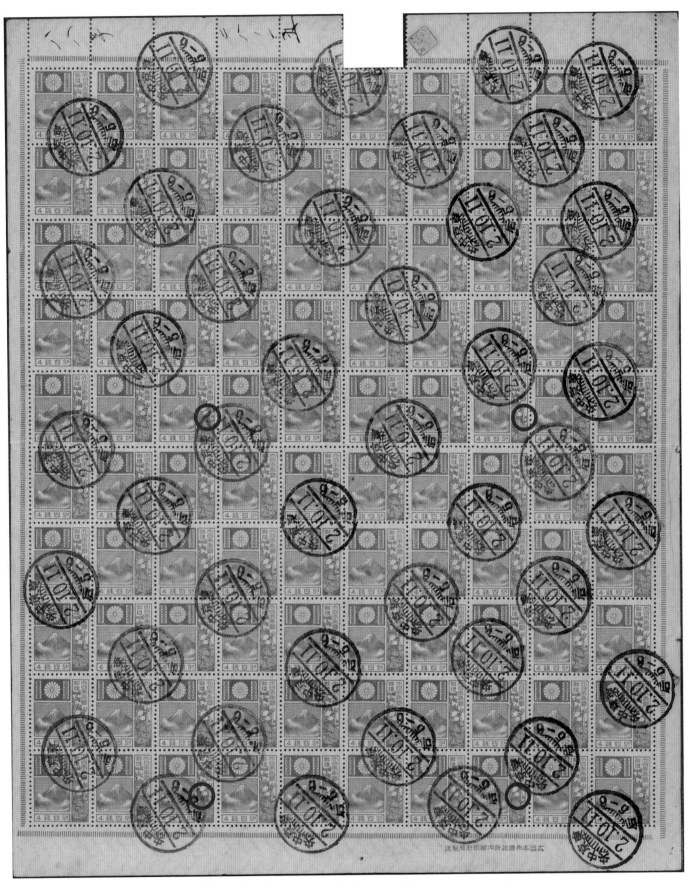

fig.8 旧版富士鹿 4 銭 1926-1929 年製造分、東京中央 2 10 11 (1927)
1922 issue 1926-1929 production, cds TOKYO CPO 2 10 11 (1927)

DIE I の印刷実用版の違い

さて、ここまで原版・二次原版に多くの紙幅を割いて説明してきました。伝統郵趣において版の分類はもっとも基本的な分類ですから、これ以外の要素は、この分類の上の乗って立つものになります。具体的には次のような要素が加わり、更に印刷ペーンの分類及び製造時期の特定が可能になります。

（１）銘版の種類

（２）銘版・罫線の位置

（３）罫線の形状

（４）トンボの形状／位置

（５）外部罫線の形状／位置

（６）刷色変更時の印刷版

（７）目打の粗さ

（８）糊引き工程

Difference of the Panes of DIE I

I've explained the original die and the Second Die in detail. The classification of dies is essential in the traditional philately, and other elements are added. Specifically, the following are additional elements to know the pane classification and printing period.

(1) imprint

(2) position of the imprint or the ruled lines

(3) shape of the ruled lines

(4) shape and position of the register marks

(5) shape and position of the outer ruled lines

(6) printing pane for the revised color

(7) roughness of perforations

(8) gumming process

（１）DIE I の銘版の種類

1922 年から 1929 年までに製造された富士鹿切手は、関東大震災や、印刷所の名称変更等により、銘版付き切手を４種類に分類可能です。この分類は容易な為、多くのカタログで掲載され、銘版を完集した収集家も多いと思います。(tbl.4, fig.10)

(1) Imprint of DIE I

The 1922 issue can be classified into four groups by imprints. The imprint changed because of events such as the Great Kanto Earthquake and the change of the name of printing bureau. It's easy to classify them, and many catalogues have the information. Some collectors may have complete imprint varieties (tbl.4 and fig.10).

	刷色 SHADE	銘版の文字数 Num of imprint	銘版 Pos. Pos. of imprint	罫線 Ruled line	発売開始時期 issue date	製造時期 Production period	トンボ register mark	糊 塗布時期 gumming *
1	緑	12	95-96	○	1922.1.1	1921-1923	ドット dot * 2	印刷後 after
2	緑	12	95-96	×	1924.-.-	1924-1925	不明 unknown	印刷前 before
3	緑	14	95-96	○	1925.-.-	1925-1926	ドット dot * 4	印刷前 before
4	緑	14	97-98-99	○	1926.-.-	1926-1929	ドット dot * 4	印刷前 before
5	橙	14	97-98-99	○	1929.9.-	1929-1930	十字 cross * 2	印刷前 before

tbl.4 富士鹿切手４銭 DIE I に見られる４種類の銘版の区分
4 imprint varieties on DIE i
* gumming after=gumming after printing, before=gumming before printing

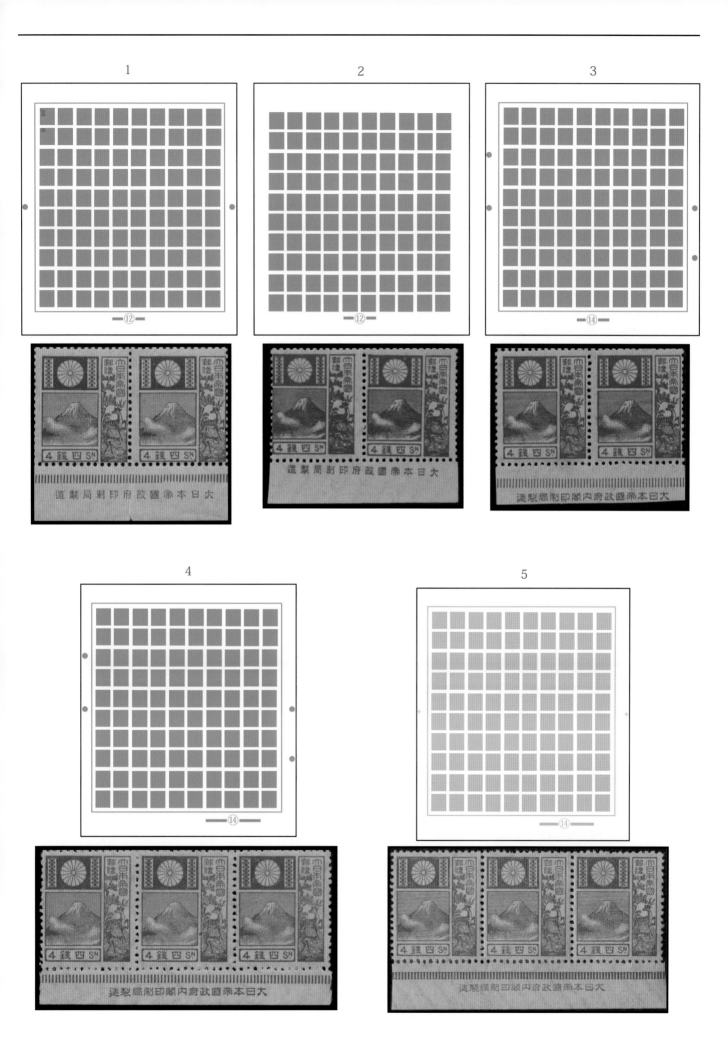

fig.10 tbl.4 で取り上げた 5 種のシート構成と銘版の違い
Differences of 5 sheets and imprints listed on the tbl.4

（2）DIE I の銘版・罫線の位置

もっとも種類の区分ができるのはあくまで 1922 年発行の旧版富士鹿切手のみで、改色切手以降は 4 番目の銘版しか存在せず、変化がありません。

そこで、銘版の種類に関わらず、製造面研究に活用したいのが、銘版や罫線の位置です。

既に説明の通り、凸版印刷で製造された富士鹿切手は、一面が 100 枚から構成されるシートを 1-4 ペーン持つ印刷版により製造されています。

二次原版の転写が終わると、罫線、銘版、トンボなどが適当に配置されるのですが、この作業は二次原版の転写ほど厳密に行われていない為、ペーンごとに位置が異なることが多く、測定結果によっては、異なるペーンで印刷された切手であることを証明することが可能です。

例えば、fig.11 に示されたマルチプルは、全て 1922 年 旧版富士鹿 4 銭の第一期製造分ですが、切手と罫線の距離、罫線と銘版の距離、切手と銘版の位置関係がそれぞれ異なっており、異なる印刷版ペーンで製造されたものであることがわかります。筆者の確認は現時点で 6 ペーンですが、まだ増えるのではないかと思います。

なお本稿では一部割愛しますが、この方法で 1930 年代前半までに製造された切手のペーン分類が可能です。

(2) Position of the Imprint and the Ruled lines of DIE I

However, imprints are helpful elements only for the 1922 issue. After the color change, all the stamps have the 4th imprint.

Then, the position of the imprint and the ruled lines help you to study the manufacturing.

As described above, the 1922-1937 issue produced through typography was manufactured from the printing plates consisting of 1 to 4 100-subject panes.

The ruled lines, imprint, and register marks were put after the transcription of the Second Die. This process wasn't so strict such as the transcription. So, the position varies on each pane, and that can verify that the stamps were produced by different printing panes.

For example, the multiples shown in fig.11 are all the first production of the 1922 issue 4 Sen. Every item has a different distance between the stamp and the ruled lines as well as between the ruled lines and the imprint. It shows that they are produced through the different printing panes. I've found six kinds of printing panes by now but expect to find more.

I've not explained all the cases, but this method makes it possible to classify the panes produced until the first half of 1930.

fig.11-1 pos.95/96

fig.11-2 pos.95/96

fig.11-3 pos.96/97 perfined

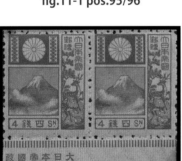

fig.11-4 pos.96/97

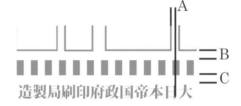

sample	A	B	C
fig.6	1.2	3.6	2.0
fig.11-1	1.2	2.7	1.7
fig.11-2	-0.4	2.7	1.5
fig.11-3	0	3.0	1.9
fig.11-4	0.4	2.7	1.5
fig.11-5	0.4	3.6	1.4

Width of each distance (mm)

fig.11-5 pos.95/96

fig.11 旧版富士鹿切手 4 銭第 1 期製造分 異なる印刷版ペーンで製造された未使用マルチプル
Different printing pane varieties of the 1921-1923 production of the 1922 4 Sen

（３）DIE I の罫線の形状

罫線は位置だけでなく、形状にも注目し、印刷ペーンの分類に役立てることが可能です。銘版と罫線の位置に変化がある以上、罫線の形状にもバラエティがあるはずなので、仮説を信じてマテリアルを探しています。

ただ、この調査の難しいところは、銘版の情報がない場合、第二期製造分の白耳を除き、製造時期の特定の決め手がなく、参考になるマテリアルの数が圧倒的に少ないことです。この為、シートや大型マルチプルに沢山触れてシェード分類に慣れ、単片でも時期区分ができる様になるのには、筆者はそれなりに時間がかかりました。

ちなみに、苦労はしたものの、シェードによる単片の分類は、説得力のある競争展示の中では最も曖昧な分類方法です。区分ノウハウはあくまで研究目的として使用し、競争展では、明確に時期を示すことができるマルチプルを主体に展示をする様に心がけています。

ところで、そのようなマテリアル探しをしている中で、これまで三回だけ罫線が断裂した異常なマテリアルに出会いました。(fig.12)

同様の罫線異常は、凸版輪転印刷（通称ゲーベル印刷）で製造されたシートで出現した事例が、田沢切手や昭和切手に確認されています。(fig.13) しかし、1922年旧版富士鹿切手は凸版平版印刷ですので、それとは状況が異なります。

これらについては『一度作成された印刷版を分解して再び組み直したもの』という説や『切手１０枚分に相当する長い罫線の代用で、バラバラに分割された罫線を埋め合わせた』という説を聞いたことがあります。

しかし、いずれの説明も信ぴょう性の高い文献の記事として未確認である為、筆者は疑問を持っています。引き続き原因を究明したいと考えています。

(3) Shape of the Ruled lines of DIE I

The shape of the ruled lines, as well as the position, will help to classify the printing panes. If the positional relationship between the imprint and the ruled lines changed, the ruled lines must vary in shape. Now I'm looking for suitable materials on this hypothesis.

This study method have a issue. Without imprint, it's not easy to classify the four printing periods. Marginal stamps without ruled lines can be classified into the second production, 1924-1925 easily, however, all the other multiples can't be classified easily. So I try to see imprinted multiples as much as possible to understand the shade deffernce, and now I can classify most of the stamps even it is a single material.

The color shade classification, which was hard to learn for me, is the vaguest method for the conviction-required competitive exhibit. So, I utilize the know-how of classification just for study and show mainly the evident printing period's materials in the competitive exhibition.

While I'm looking for the materials, I only found irregular broken ruled lines three times (fig.12).

Similar irregular ruled lines are also found in rotary press manufactured the Tazawa and the Showa issue. However, the 1922 issue was only produced by flat press, but rotary press.

Someone says, "They disassembled the produced printing pane and reassembled them again," or "They used the broken several ruled lines to substitute for long-ruled lines."

Whichever opinion is still in the study of the credibility, and I doubt them. I continue investigating the cause.

fig.12 縦方向の罫線が断裂している事例
Irregular vertical break of ruled lines

fig.13 罫線異常の田沢切手（輪転印刷）
Break ruled lines example, rotary press

（４）DIE I のトンボ類の形状／位置

切手シートに見られるトンボ類の目的には、複数シートを同時に目打穿孔する為の固定（ドット）と、裁断時にも利用できるガイド（十字等）の二つがあります。

手持ちの DIE I から作られた切手に存在が確認できたのは前者のみで、後者はありませんでした。fig.14 の様にぼてっとした形状ですが、それでもポジション特定の役に立ちます。

ところが、画像でしか残っていない改色切手（旧版）の未使用シートには十字ガイドが見られるものが確認されています。(fig.15) 1922 年に発行された富士鹿切手に同様のトンボは今のところ見つかっていないので、両者は別の印刷版で製造されたことになりますが、調査数不足で結論を出すに至りません。

(4) Shape and Position of the Register marks of DIE I

The register marks on the printing pane had two objects: to fix the sheets to perforate correctly at once (dots), and also for guidance in cutting (cross, etc.).

I have no sheets with crossed register marks of the 1922 issue and the 1929 issue DIE I. Every sheets have a prominent dot, as you see in fig.14, but it helps to identify a position.

Meanwhile, the photo illustration of a sheet of the 1929 issue DIE I 20 Sen with crossed register marks remains now (fig.15). As no sheets of the former issue with crossed register marks found by now, it means both sheets were printed from the different printing plates. We need more investigation.

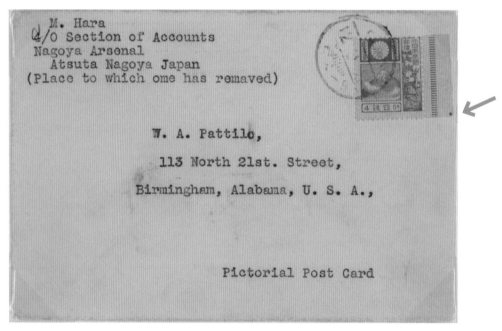

fig.14 pos.50 一枚貼り葉書使用例
名古屋→米国、12.3.2-（1923）印刷物（絵葉書）
Single franking of the Pos.50, post card usage
NAGOYA to the US, Mar. ? 1923 pictorial post card as printed matter

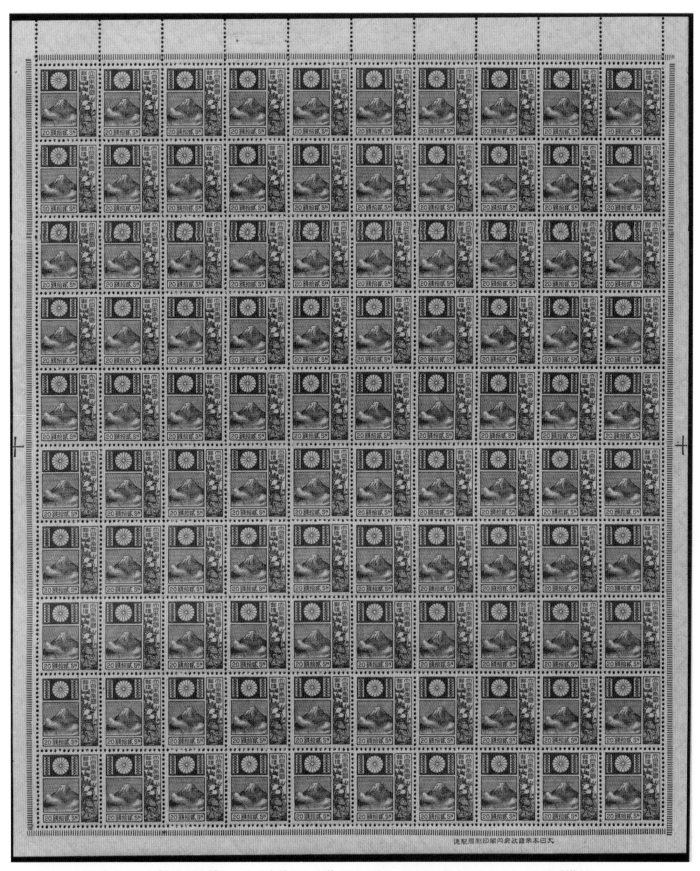

fig.15 1929 年シリーズ DIE I 20 銭シート (未使用) 香港 JOHN BULL 2007.3 月オークション 12 万香港ドル ->UNSOLD
Unsold unused sheet of 1929 20 Sen DIE I, which were consigned to JOHN BULL March sale, 2007, 120,000 HKD

（5）DIE I の外部罫線の形状／位置

外部罫線とは、人によって呼称が変わる、耳紙上に見られる色付きの棒の事を言います。色検査用のカラーマークだという人もいますが、印刷時に用紙を抑える為の、外部罫線ではないかというのが私の考えです。

1922年シリーズ第1期製造分の在庫と印刷版が関東大震災で失われると、耳紙から罫線が消え、いわゆる「白耳」と言われるマージンになります。しかし、罫線の役割は完成したシートにおける見た目よりも、印刷版における切手部分への過剰な圧力の除去でしたので、代わりとなる罫線が必要でした。このため、通常は印刷後に裁断される部分に罫線が置かれ、印刷されていたことが、裁断不完全シートのマルチプル (fig.16) から判明しています。

1925年4月頃には「白耳」だった現行切手の印刷版は、罫線ありに戻りますので、この外側の罫線は不要になるはずです。しかし1925年から1945年にかけて製造された切手シリーズの内、第一次昭和切手を除く全てのシリーズで、確認事例があります。したがって、製造上何らかの理由であったほうが良いと判断されたことがうかがえます。(fig.17)

fig.16, 17 で見られる通り、外部罫線の形状は、大きく分けて2種類で、100面シートの周囲に（1）平行な罫線（1本）と（2）垂直な罫線（1～3本、太さも様々）です。

同時代に製造された田沢切手半銭（旧大正毛紙）には、pos.6 の上部マージンに2本の縦棒、また pos.10/100 の右マージンに10の縦棒が未裁断のまま残っているシートがありますので、二つの外部罫線が共に一つの印刷ペーンに存在する事例も確認済みです。(fig.18)

ちなみに「（1）平行な罫線」は、100面シートの縦方向の左右に見られることが多いのですが、上部や下部に見られる事例も確認されています。(fig.19)

(5) Shape and Position of the Outer Ruled Lines of DIE I

Outer ruled lines is color bars on the margin, which are also called color mark. However, I think it was printed to hold the paper at the printing procedure, and not for inspecting colors.

The Great Kanto Earthquake had lost the kept stamps of the 1st printing of the 1921-1923 production and the printing panes, the newly reproduced printing plates had no ruled lines on the margin, so thus sheets are called "white margin." However, the ruled lines didn't only worked for good looking but removing the excess pressure put on the printing plates or hold papers at the printing procedure. So, they needed other ruled lines for this role. Some multiples of the incompletely cut panes (fig.16) show that the panes have the outer ruled lines on a part to be cut after printing.

Since the printing panes with "white margin" was replaced by new printing plates with ruled lines around in April 1925, the outer ruled lines shouldn't have been needed. But the outer ruled lines are recorded in all the issue between 1925 to 1945, except the 1937-1940 issue. It's possible to have left them unremoved for some usefulness in the printing process (fig.17).

You can see by fig.16 and 17 that the shape of the outer ruled lines can be generally classified into two types: (1) one parallel line to the frame of a 100-subject pane and (2) one to three vertical lines (the width varies).

The Tazawa 1/2 Sen (DIE 1 Granite Paper) manufactured in the same period includes the panes with both of two vertical lines on the margin above pos.6 and ten unremoved vertical lines on the right margin between pos.10 and 100. (fig.18).

"(1) parallel ruled lines" are often seen vertically on both sides of a 100-subject pane, but a few panes have them horizontally at the top or bottom (fig.19).

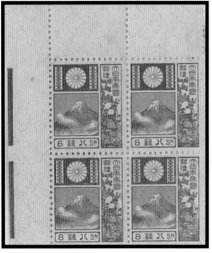

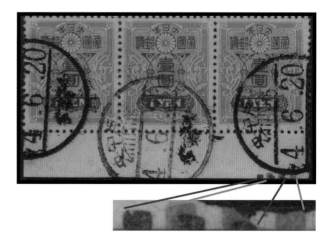

fig.16 裁断されなかった外部罫線（周囲に平行、周囲に垂直2本、周囲に垂直3本＊2色）
Irregular outer ruled lines which were normally cut off

fig.17-1 4銭第3期に見られる外部罫線 pos.3/5
1922 issue 4 Sen, 1925-1926 production, Pos.3/5

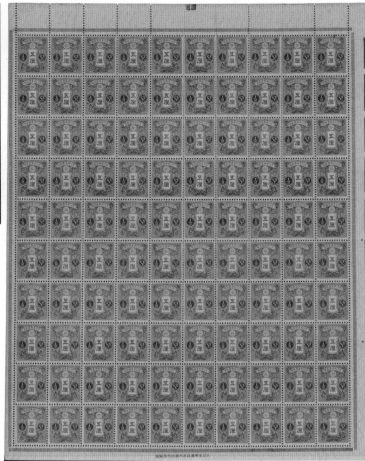

fig.18 罫線に平行の外側罫線と垂直の外側罫線が共に存在するシート
A sheet with both 2 kinds of outer ruled lines

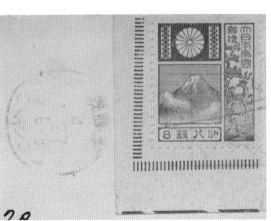

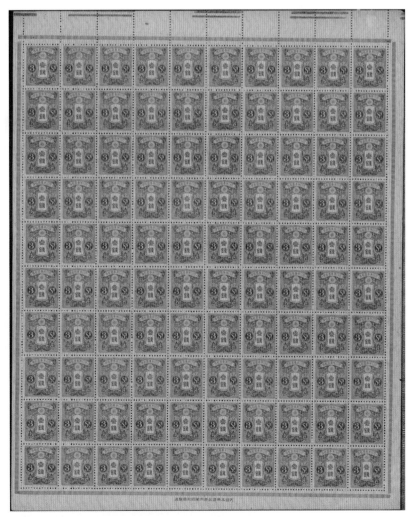

fig.19 様々な外部罫線 , various outer ruled lines varieties

（6）DIE I の刷色変更時の印刷版

1922 年に発行された旧版富士鹿切手は、1925 年 10 月 1 日に外信料金が 10 銭、6 銭、2 銭（書状、葉書、印刷物の順）に改定されると、その本来の役目を失います。

翌年の 1926 年 7 月 5 日に、1926 年シリーズ（一般カタログでの呼称は「風景切手 第一次（平面版）」）が発行された後、UPU カラーであり続けた 20 銭青、8 銭赤、4 銭緑の色が変えられたのは更に三年後の 1929 年になってのことでした。

これが富士鹿改色切手ですが、一般カタログに掲載されて有名な通り、改色切手には旧版と新版が存在し、前者の製造期間が圧倒的に短いことで有名です。DIE I について解説している本項では旧版についてのみ解説します。

富士鹿改色切手(旧版)の印刷実用版についてはよくわかっていません。以前まで筆者は、直前まで製造されていた旧版富士鹿切手の印刷実用版をそのまま用いて、インクの色だけ変えたと考えていましたが、最近は新たに整版された可能性が高いと考えています。最大の理由は、改色富士鹿切手の唯一のシートとして画像が現存する 20 銭未使用シート (fig.15)に、旧版富士鹿切手では未確認のトンボがあることです。

平行して進めている印刷ペーンの罫線・銘版の位置や形状の調査 (fig.20) でも、両者が一致する事例が依然として見つかっていないこともこの仮説を裏付けています。

もっとも、この仮説は、両者が一致する事例が一点ないし数点見つかれば覆る仮説ですので、今後も事例調査を進めたいと思います。

なお、改色切手（旧版）が新たに用意された印刷実用版を用いて製造された場合に、二次原版の構成要素数が何枚であるかも気になるところですが、現時点では 10 面か 25 面か特定できるマテリアルを未確認です。

(6) Printing Pane for the Revised Color of DIE I

The 1922 issue lost its initial role when the international postage rate was changed to 10 Sen, 6 Sen, and 2 Sen (postage of letter, postcard, and printed matter, in this order) on October 1, 1925.

The next year, on July 5, 1926, the 1926 issue (generally, "Landscape Stamp, the 1st printing, flat press" in traditional catalogue) was issued. Three years after then in 1929, the colors of the 1922 issue: 20 Sen blue, 8 Sen red, and 4 Sen green was changed.

They are the 1929 issue. It's famous that there are old DIE and new DIE in the issue, and the former of which were issued only for a short period. This chapter only shows DIE I.

The printing plates of the 1929 issue, DIE I is still under investigation. It used to be thought that it was printed only by changing ink colors with the same printing plates used for the 1926-1929 production of the 1922 issue, however, this hypothesis was denied by the photo illustration of 1929 issue, DIE I 20 Sen, which has crossed register marks not recorded on the 1922 issue. (fig.15)

I'm investigating the position and shape of the ruled lines and etc (fig.20) , but haven't found any case that both dies were the same features. It also deny old hypothesis.

But just one matched material showing the same features come will quickly change the situation, so I'll keep investigating the cases.

It is also unknown how many subjects consists the second DIE, but None of the materials showing whether 10 or 25 has been found by now.

fig.20-1 Pos.87/99

fig.20-2 Pos.87/100

fig.20-3 Pos.97/100

fig.20-4 pos. 97/99

fig.20-5 Pos.87/99

fig.20-6 pos. 98

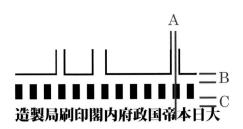

sample	A	B	C
fig.20-1	1.0	2.8	1.0
fig.20-2	-3.8	3.6	2.4
fig.20-3	0.5	3.8	0.8
fig.20-4	1.5	3.8	1.0
fig.20-5	-0.5	2.8	1.7
fig.20-6	-1.8	3.5	1.0

Width of each distance (mm)

fig.20 旧版富士鹿第４期と改色富士鹿切手（旧版）の印刷ペーンの調査
A Study on printing panes, 1922 issue 4 Sen, 1926-1929 production and 1929 issue 4 Sen

（7）DIE I の目打の粗さ

1922年発行の富士鹿切手の目打穴を見ると、細穴の目打を垣間見ることができます。関東大震災の前後に関わらず出現しており、目打穿孔機の整備不良で一括りにされてきましたが、もう少し頭を使っても良さそうですので、調査してみたいと思います（fig.A-1）。

（8）DIE I の糊引き工程

1922年発行の富士鹿切手の上耳の糊引きを調べると、1921-1923年製造分で、上耳の半分程度までしか引かれていない（fig.A-2）反面、1924-1925年製造分では、上耳全てに引かれている（fig.A-3）様です。今のところ反例が発見されていない為、前者では印刷および用紙裁断後に糊が引かれ、後者では印刷前に糊が引かれたということが推定できます。

(7) Roughness of Perforations of DIE I

Rough perf. with small holes can be seen in the 1922 issue, and it appears regardless of the production period; before and after the Great Kanto Earthquake. It has long been thought that poor maintenance of the perf. machine caused this, however, this hypothesis is too hasty. I'd like to collect materials and study more, fig. A-1.

(8) Gumming Process of DIE I

Seeing the back of the 1922 issue found that only half of the upper margin is gummed in the 1921-1923 production, fig.A-2, while all the area is gummed in the 1924-1925 production, fig.A-3. No counterexample has been found, and you can see the cutting sheets process precedes gumming process in the former production and the gumming process precedes printing process in the latter production.

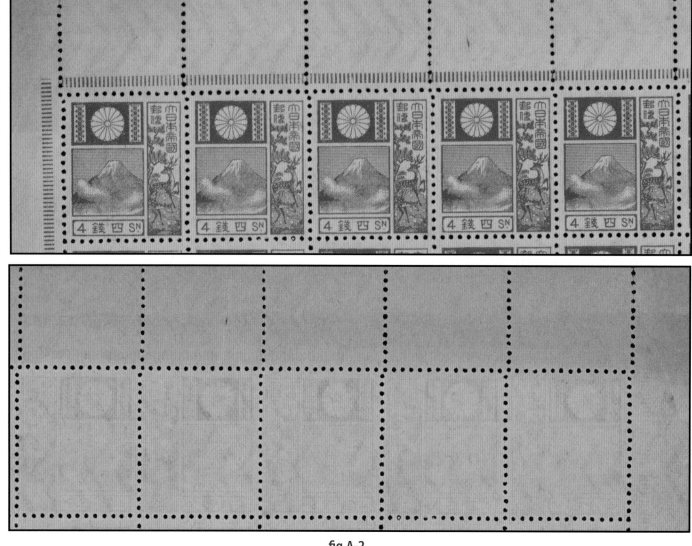

fig.A-2
第 1 期製造分（1921-1923）では裏糊が上部マージンの半分程度までしか引かれていない。
裁断後に糊が引かれたことが推定される。
1921-1923 production, upper margin is half gummed, which tells the cutting sheets process precedes gumming process.

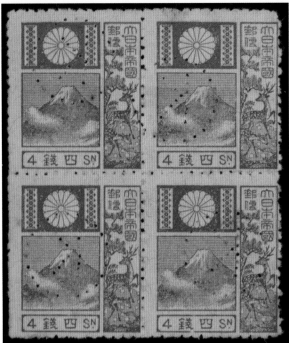

fig.A-1
1922年シリーズ 4銭、8銭の目打の粗いマテリアル、
the 1922 issue 4 Sen, 8 Sen with rough perf.

Original Die, DIE II

The Printing Bureau bought a rotary press of German Goebel Printing in August 1924. This machine was a rotary press, which enabled doing all the manufacturing processes of the maximum bi-color typography stamps continuously: printing, perforating and cutting sheets. The paper was rolled and pre-guming, it means they didn't need another gumming process.

The stamp production by this new machine started in February 1926. The order of appearance until 1935 was 3 Sen (1926, fig.21), 30 Sen, 50 Sen (1929), 1 Yen (1930), and 1 1/2 Sen (1931) of the Tazawa; the Landscape 2 Sen (1932), and the Tazawa 5 Rin (1935).

When the die for the flat press was used in the rotary press, the stamps' long side would become longer than the one produced in flat press. It was caused when the Second Die was installed to rolling copper plates in rotary press.

The printed size, 19.0 x 22.5 mm, of the 1922-1937 issue and the Tazawa was possibly changed to about 19.0 x 23.0 mm by stretching after the installation in the rotary press. It was not adequate to have correct perforations.

Therefore, after 1926, they made a new DIE II, which was slightly smaller than the existing one by around 0.5mm. "Japanese Stamp Directory vol. VI" (JPS,1980) says on p.8 that DIE II was "made by the photomechanical process, which had already been up to practical use then." Moreover, the same article analyzes, "the (Tazawa) New Die was prepared by making a clean proof with the 2nd die of the old issue with the face value, enlarging or fixing it if necessary, and then reducing it."

fig.A-3
第２期製造分（1924-1925）には裏糊が際まで引かれており、
印刷前に糊が引かれたことが推定される。
1924-1925 production is gummed by the edge,
which tells the gumming process precedes printing process.

原版 DIE II

　1924 年（大正 1 3 年）8 月末に、印刷局はドイツのゲーベル社から、凸版輪転印刷機を購入しました。この機械の特徴は、単に輪転印刷であることにとどまらず、2 色刷りまでの凸版切手について、印刷・目打穿孔・裁断が一台の機械の中の連続した工程で行われることにありました。用紙も予め糊引きされたロールペーパーを使用するため、糊引き工程が別途行われている訳ではありません。

　この新型機による切手の製造は、1926 年（大正 1 5 年）2 月に開始されました。1935 年（昭和 10 年）までの出現順は、3 銭（1926= 出現年 , fig.21）、30 銭、50 銭（1929）、1 円 (1930)、1 銭 5 厘 (1931, 以上田沢切手)、風景 2 銭（1932）、田沢 5 厘 (1935) でした。

　ところで平版印刷機で製造されていた切手を、同じDIE を使用して、凸版輪転印刷機で製造すると、印面寸法の長辺が伸びる状況が見られます。この原因は一般的には第二次原版を輪転印刷機に取り付ける時の湾曲で寸法が伸びる為と説明されています。

　富士鹿切手や田沢切手の DIE の印面寸法 19.0 x 22.5 mm は、輪転印刷機に取り付け、縦方向に寸法が伸びてしまうと、19.0 x 23.0 mm 前後になってしまうと考えられます。この印面寸法はそのままでは正常な目打穿孔が期待できない長さです。

　そこで、1926 年（大正 15 年）以降、既存の印面寸法をわずかに（約 0.5mm）縮小した、新しい DIE（「DIE II」と呼びます）が用意されました。『日本切手名鑑 第 6 巻』（JPS,1980）P.8 では、DIE II は、『当時すでに実用の域に達していた写真製版法によって作られた。』と記載があります。同記事ではさらに、『(田沢) 新版切手は、旧版切手のすでに額面の入った第 2 次原版の清刷りをとり、必要があれば拡大・修正して、額面ごとに写真縮写を行ない、新しい原版を作ったようだ。』と分析しています。

　このようにして作られた DIE II の印面寸法は、平版印刷機に取り付けると、18.5 x 22.0 mm となり、また輪転印刷機に取り付けられると、18.5 x 22.5 mm になりました。田沢切手の中でも 1 銭、5 銭、7 銭、13 銭、25 銭は輪転印刷機で製造されず、平版印刷機で製造され続けましたが、全ての DIE II が縮小されたため、1930 年（昭和 5 年）以降、DIE II を元にした印刷版で製造された、印面寸法が縦横共小さくなった切手（新大正毛紙）が登場しました。

　全て凸版平面版印刷機で製造されたと考えられる、改色富士鹿切手でも、田沢切手同様に、印面寸法が縦横共小さくなった新版が 1930 年より登場します。

The printed size of DIE II through flat pree became 18.5 x 22.0 mm, and 18.5 x 22.5 mm through rotary press. 1 Sen, 5 Sen, 7 Sen, 13 Sen, and 25 Sen of the Tazawa issue were not produced by rotary press but flat press. Since all the size of stamps produced through DIE II were reduced regardless if they were printed through rotary press or flat press, they were classified to new production.

The 1929 issue was produced through flat press, however, DIE II was introduced in 1930.

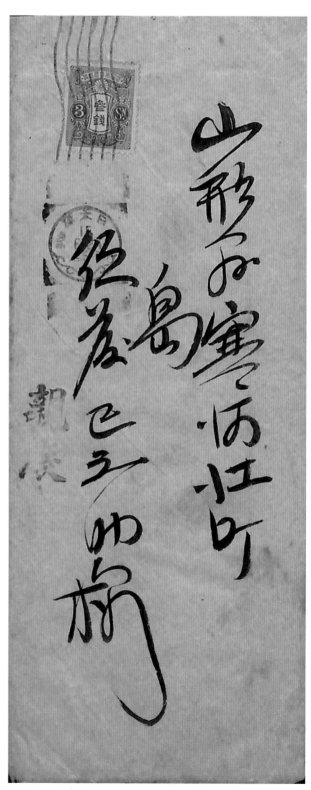

fig.21 田沢・新毛 3 銭輪転版 初期使用
日本橋 機械印 15.6.24 発行 5 ヶ月目
Early usage of a stamp manufactured by rotary press, Nihonbashi 6 24 (1926)

改色富士鹿切手（新版）は、郵便局では 100 面シートで販売されますが、凸版平面版印刷機では、1 面から 4 面掛けまでの印刷版に対応できることがわかっています。もっとも記録は残っていないため、どの時期にどのような印刷版が使用されたかは不明です。

二次原版についても、DIE I から製造された印刷版に出現する「キーポジション Type 8」の様な目立つ変種は未発見の為、おそらく 25 面だろう、程度しかわかりません。

1930-1931 年より発売の始まったこれらの切手は、1937 年までの 6-7 年間製造・使用されたため、製造数も多く、シート残存量も DIE I のそれより多いと予想されますが、現時点の確認数は 4 銭シート 4 点に留まります。

しかしながら、罫線の形状の調査から 1937 年に発行された昭和毛紙、昭和白紙は改色富士鹿切手（新版）最後期の印刷実用版を用いて製造されたことがわかっています (fig.22)。刷色の違いはありますが、3 つのカタログのメインナンバーをまとめて調査することで、DIE II 後期の印刷実用版を解明することが可能です。

なお、銘版が一種類しか存在しない為、この時期の切手の印刷実用版の区別には、下記を使います。

（1）罫線・トンボの形状／位置

（2）裏糊

（3）銘版・罫線の位置

The 1929 issue DIE II was sold at the post office at a sheet of 100 subjects, and it is known that typography flat press could print from one to four panes of 100 copies at once. Since no record remains, it is unknown which kinds of printing panes were used in which period.

The details of the Second Die are also unknown. Since I haven't found an appealing variety such as "Key Position Type 8" of DIE I, I just suppose it consists of 25 subjects.

Since these stamps issued from 1930 or 1931 were manufactured and used for six or seven years to 1937, I expect a large amount of production and more surviving sheets than DIE I. But the number I've confirmed for sheets of 4 Sen is only 4 in this 7 years and no sheets for the other two denomination stamps.

However, the investigation of the shape of ruled lines shows the Showa Granite Paper and the Showa White Paper, which were issued in 1937, were produced from the same pane of the latest printing period of the 1929 issue DIE II (fig.22). Although they have different colors, it will be possible to clarify the printing panes of the late printing period of DIE II.

They have only one imprint, so I use the following elements to classify this period's printing panes.

(1) shape and position of ruled lines and register marks

(2) gum

(3) position of imprint and ruled lines

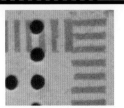

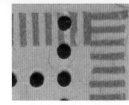

fig.22 同一形状の罫線を持つ、新版改色、昭和毛紙と昭和白紙の富士鹿切手 pos.10, All of them are with the same ruled lines
①上部マージンに印刷される縦罫線の右端：太めで右マージンに印刷されている横罫線に接する。
②上部マージンに印刷される縦罫線の右から 2 本目：株がえぐられたようになっており、下に行くにつれて、細い。
③右部マージンに印刷される横罫線の上から 5 本目：他よりも長さが短い。

　改色富士鹿切手のシートはほとんど残っていませんが、筆者が持つ未使用 (fig.25) と使用済 (fig.24) のシートを見比べたところ、罫線・トンボの形状と位置が異なるものでした。

　1922 年に発行された旧版富士鹿切手には裁断の目安となる十字トンボがなかった為、シートの裁断は曲がることが多く、1930 年前後から十字トンボが印刷されるようになりました。

　これにより、印刷ペーンを区分する手段がまた一つ植えました。つまり、トンボの種類や位置が異なれば、印刷ペーンも同一ではないということです。(tbl.4 参照)

　また、1937 年に発行された昭和毛紙、昭和白紙を調べると、いずれも fig.25 と同じ特徴を持つことがわかりました。(fig.26) この点からも、昭和白紙も含めてこれらの切手が同一の印刷ペーンにより製造されたシートであることが分かります。

　なお、DIE II で製造された切手で、罫線異常が一例だけ見つかりましたので、ご紹介いたします。(fig.23)

　これまでの説明でお分かりいただけると思いますが、富士鹿切手の縦方向の罫線は原則として、10 枚分の長さがあり、途中で断裂しませんが、pos.10/20 の位置で断裂しているものです。ただこれは、元々連続していた罫線の一部分が欠けたものであることが分かりました。元の印刷版は、「2 箇所にトンボ」のシート (fig.24) と同一ですので、初期の印刷版で刷られたと言えます。

Not many sheets of the 1929 issue survive now. When I compared the unused (fig.25) and the used (fig.24) sheets in my collection, they have different shapes and positions of the ruled lines and the register marks.

Since the 1922 issue didn't have any crossed register marks for cutting correctly, sheets were cut crooked. Then, around 1930, they started printing crossed register marks.

Now we've got another element to classify printing panes. It means if the position or shape of register marks are different, they are not produced by the same printing pane. (see tbl.4)

Both the Granite Paper and the White Paper, used for the 1937 issue, have the same feature of fig.25 (fig.26). It also shows the same printing plate produced these sheets.

Here I show an irregular ruled lines found in DIE II (fig.23).

I have already described the vertical ruled lines on the 1922-1937 issue measure principally for ten subjects without interruption. But that pane has an interruption between pos.10 and 20. It initially had a partly-lacked ruled line. The original pane is the same one of "two register marks" (fig.24), so it happened on the initial printing plates.

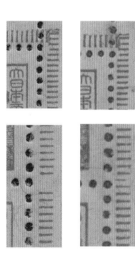

fig.23
罫線の一部が欠損した例（左）と元の状態（右）
A part of ruled lines missing on the left, and original one on the right

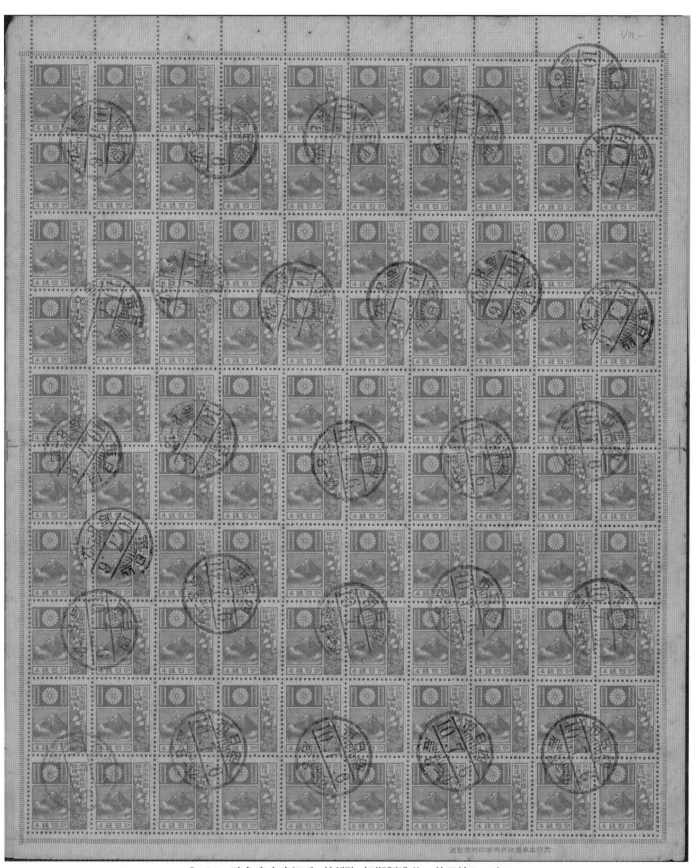

fig.24　改色富士鹿切手（新版）初期製造分　使用済シート
fig.27-1 の銘版を持ち、pos.41/51 と pos.50/60 の二箇所にトンボが印刷されている
1929-1937 issue 4 Sen DIE II, sheet, same imprint having imprint of the fig.27-1, with 2 register marks

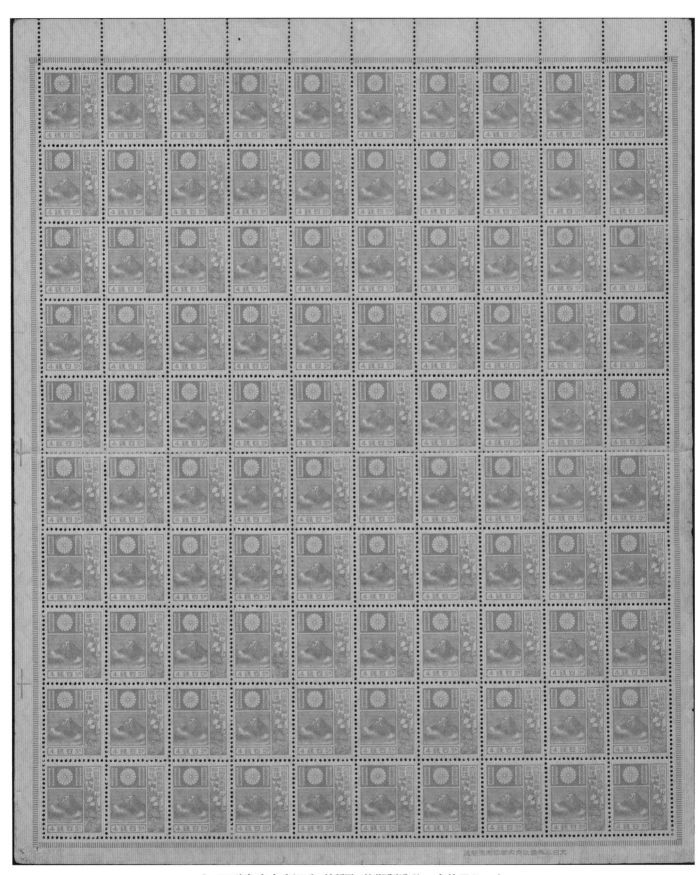

fig.25 改色富士鹿切手（新版）後期製造分　未使用シート
fig.27-2 の銘版を持ち、pos.20/30, 41/51, 50/60, 71/81 の 4 箇所にトンボが印刷されている
1929-1937 issue 4 Sen DIE II, sheet, same imprint having imprint of the fig.27-2, with 4 register marks

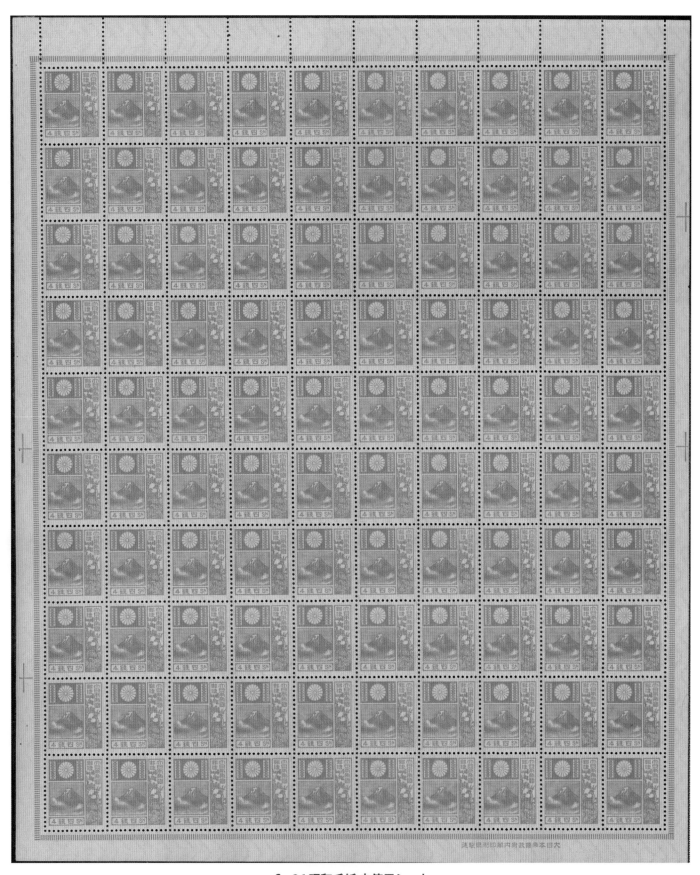

fig.26 昭和毛紙 未使用シート
fig.27-2 の銘版を持地、pos.20/30, 41/51, 50/60, 71/81 の 4 箇所にトンボが印刷されている
1937 issue 4 Sen granite paper, sheet, same imprint having imprint of the fig.27-2, with 4 register marks

DIE II の裏糊

ところで、改色富士鹿切手が製造されていた 1930 年代前半には、切手の裏面の糊の改善が図られ、デキストリン糊が登場しました。

デキストリン糊の最大の特徴はその光沢で、完全な状態の場合、光に照らすと光沢が全面に表れます。ただスキャンでは再現しない為、3 枚ほど写真撮影しましたので参考にしてください。いずれも左がアラビア糊で、右がデキストリン糊です。

Gum of DIE II

In early 1930's, when the 1929 issue was on sale, the gum was improved, and dextrin gum replaced arabic gum which were used until then.

Dextrin gum's feature is its gloss, and mint dextrin gum shows its gloss on the entire back of the stamp. It does not show by scanning the back of the stamp, so I try to show three photos taken from slightly different angles. The left is Arabic gum and the right is Dextrin gum.

Photo.1 アラビア糊（左）とデキストリン糊（右） The left is Arabic gum and the right is Dextrin gum.

改色富士鹿切手（新版）4 銭は、1931 年から 1936 年まで 6 年間製造された為、複数の印刷ペーンがあると考えられます。この仮説のもと、手元の銘版部分のマルチプルを調査したところ、まず 2 種類が確認できました。

この 2 つの印刷ペーンが同一実用版の異なるペーンなのかは銘版部分だけではわかりませんが、先述したトンボの位置により、fig.24 の一部である fig.27-1 は初期の印刷版であり、fig.25 の一部である fig.27-2 は後期の印刷版である事がわかります。

なお 1937 年に製造された、昭和毛紙と昭和白紙の 4 銭富士鹿切手について銘版部分の調査をしたところ、改色富士鹿切手（新版）後期の fig.25 に見られるパターン以外は一例も確認できませんでした。1932-4 年頃以降は、罫線や銘版の印刷版への設置精度が高まったと考えられ、それ以降、ペーン分類の材料にするのが難しくなります。

Since the 1929 issue DIE II 4 Sen was produced for six years from 1931 to 1936, it possibly has more than one printing plates. I've done a hypothetical investigation of imprint on the multiples in my collection, and I've found two kinds of printing panes.

It is not clear only by comapring imprints. But as they are part of sheets, fig.27-1, which was a part of fig.24, was on a printing pane of early production, and fig.27-2, which was of fig.25, was late period with information of register marks .

I also investigate imprints of the 1937 issue, but all the items I've found were the same as fig.25 regardless if they were printed on granite paper or white paper. I thought the position of ruled lines or imprints on a printing pane became accurate from 1932-4, and it is difficult to classify the pane by the position after then.

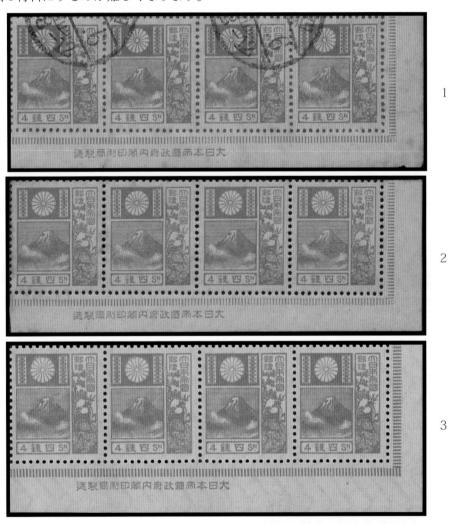

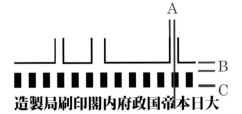

sample	A	B	C
fig.23-1	0	3.0	1.2
fig.23-2	1.0	3.4	1.2
fig.23-3	1.0	3.4	1.2

Width of each distance (mm)

fig.27 改色富士鹿切手（新版）、昭和毛紙の印刷ペーンの調査
A research of printing panes of 1929-1937 issue DIE II and 1937 issue granite paper

8 銭

8 Sen

4 銭切手を例に製造面の解説および分析方法を明らかにしましたので、残り 2 額面についてはトピックスのみ取り上げます。

I have written the details of manufacturing and the analyzing method in the long chapter of 4 Sen, so I just show some topics of the other two face values.

富士鹿切手の最難関

The Most Challenging Stamp of the 1922-1937 issue

富士鹿切手の最難関は 1922 年に発行された 8 銭赤の第 1 期製造分で、筆者も長年に渡り収集していますが、銘版付きマルチプル (fig.29-1) の入手が数点にとどまります。初期使用例も非常に少ないのですが、これらの原因は、逓信省の発売政策と関東大震災による在庫および印刷版の消失にあります。

逓信省は 1922 年 1 月 1 日の富士鹿切手の発行に伴う売り出しを行わず、旧切手となった田沢切手が完売し次第、富士鹿切手の販売を開始するよう各局に指示しました。この為、フィラテリック なものも含めても 1922 年前半の使用例 (fig.29-2) は少なく、数えるほどしか存在しません。

このような発売政策に加えて、翌 1923 年 9 月 1 日に起きた関東大震災により、富士鹿切手は、逓信省の保管在庫全てと印刷実用版を焼失した為、未使用も使用例も少ないものとなりました。そして 3 額面の中でも需要の最も少なかった 8 銭は格段に少ないものとなりました。

The most challenging stamp of all the 1922-1937 issue is the first printing of 8 Sen red, issued in 1922. I have collected it for years but only got a few imprinted multiples(fig.29-1). The usages in the early printing period are also uncommon. That is a consequence of the Ministry of Communications' sale policy and the destruction of the stamps in stock and printing pane by the Great Kanto Earthquake.

The Ministry of Communications didn't start selling the 1922 issue on the first day, on January 1, 1922, and ordered the post offices to sell them after the former 1914 issue was sold out. So, there are not many used items, even philatelic ones, in the first half of 1922 (fig.29-2).

Beside this selling policy, the Great Kanto Earth Quake in the next year, on September 1, 1923, burnt off all the stocked stamps and the printing plates. Regardless if it is unused or used, all the 1921-1923 production remains small in number, and 8 Sen among them are especially scarce because 8 Sen denomination stamps were in the least demand in three face values.

fig.29-1 第 1 期製造分 pos.85/96
1922 issue 1921-1923 production

fig.29-2 第 1 期製造分 単貼り最初期使用例
The earliest single farnking of the 1922 issue 1921-1923 production

8銭という額面の切手の需要が少なかったこともあり、DIE I を使い製造が再開された後の第2期以降の製造分も決して多くなく、比較的長期間製造された第4期でも収集には苦労させられます。

その中でもトピックスとなるのは、1924-1925年製造分で、銘版付きブロックの入手までは比較的簡単ですが、それ以外のポジションや使用例を示すには、耳紙付きのマテリアルが求められる為、難易度が上がります。耳紙付きの完全な使用例はここに示した以外にはほとんどないのではないでしょうか。(fig.30)

Since the demand for 8 Sen stamps was weak, usages of the 2nd printing period and after then were also uncommon, The 1926-1929 production was the most long produced stamps, however, it's not easy to make a good exhibition leaves even with them.

The 1924-1925 production is the most difficult among them. Imprinted block is not so difficult, however, if you'd like to show the other materials such as marginal stamps of the other position or marginal stamp franking covers are very difficult. two usages are shown here, but I've not found any other entire usages. (fig.30)

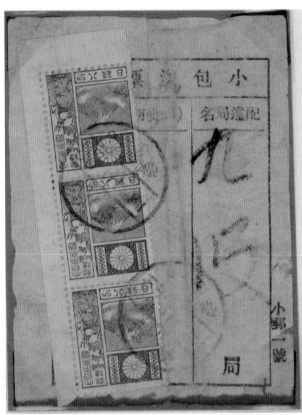

fig.30-1 第2期製造分 白耳付き 3枚ストリップ貼り使用例
麹町 1410 23 (1925) 小包送票 (600 匁迄 24 銭)
1924-1925 production with no ruled lines on the margin,
po.80/100 strip of three, KOJIMACHI 10 23 (1925) Parcel tag

fig.30-2 第2期製造分 白耳付きコーナー
Three coner blocks of the 1924-1925 production of the 8 Sen

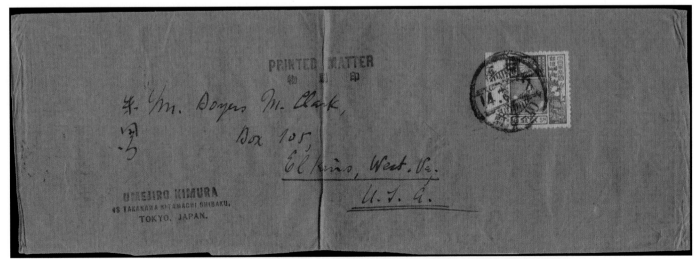

fig.30-2 第2期製造分 白耳付き貼り使用例 白金 14 5 2 (1925) 米国宛印刷物重量便
1924-1925 production with no ruled lines on the margin, SHIROGANE to the US, 5 2 (1925) double weight printed matter

　８銭切手は、1929 年に改色切手として発行された時点では、DIE Iから作られた印刷実用版で製造されたシートで発売されましたが、翌 1930 年前半には、DIE II から作られた印刷実用版で製造されたシートに切り替わっています。

　このように発売期間が半年程度と短く、また他の２額面よりも需要が少ない額面であったこともあり、旧版改色８銭切手は、カタログ価格が高いことで有名です。なおマルチプルも少なく、最大が 10 枚ブロック（fig.31-3）で１点のみ、次が６枚ブロックで 4-5 点、その次が５枚ストリップで 1-2 点、田型も 2-3 点でしょう。３枚ストリップになるとだいぶ増えますが、それでも 10-20 点でしょうか。

　この切手の pos.99 には有名な定常変種があります（fig.31-1）。それに加えて４銭切手同様の罫線・銘版の分析を用いると、現時点で少なくとも３種類の印刷ペーンを確認することができます。(fig.31-2)

　ちなみに定常変種のある銘版付きストリップも珍しいのですが、定常変種のない銘版付きストリップは２点しか確認されておらず、そちらの方が珍しいとされています。

The 1929 issue DIE I 8 Sen

When the Color Changed 8 Sen was issued in 1929, the sheet was made from DIE I, and the die changed to DIE II in the first half of 1930.

The 1929 issue DIE I 8 Sen is a high-priced stamp in the catalogue because of the short period of a sale, about half a year, and less demand than the other two face values. Multiples of 8 Sen are also scarce, a block of 10 is only one knwon, at fig.31-3, block of six 4-5 known, strip of five 1-2 knwon, and a block of four 2-3 known. A strip of three survives more, but 10-20 at most.

It has a famous constant PF of pos.99 (fig.31-1). Additional analysis of the ruled lines and the imprint like 4 Sen will help you see three kinds of the printing panes at least (fig.31-2).

The number of pos.99 having PF is around 10 examples, while pos.99 <u>without</u> PF is only two.

fig.31-1　改色富士鹿切手（旧版）８銭 の一部のシートの Pos.99 に見られる定常変種（右は通常のもの）
A plate flaw of the pos. 99 of a part of sheets of the 1929 issue DIE II 8 Sen （The right is normal one）

1

2

3

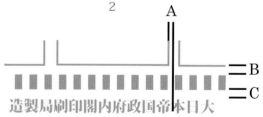

sample	A	B	C	定常変種 PF
fig.31-1	0.1	4.2	1.0	without PF
fig.31-2	0.1	3.6	1.2	without PF
fig.31-3	1.0	4.2	0.8	with PF at pos.99

Width of each distance（mm）

fig.31-2　改色富士鹿切手（旧版）８銭の印刷ペーンの調査
A study of printing panes of 1929 issue 8 Sen DIE I

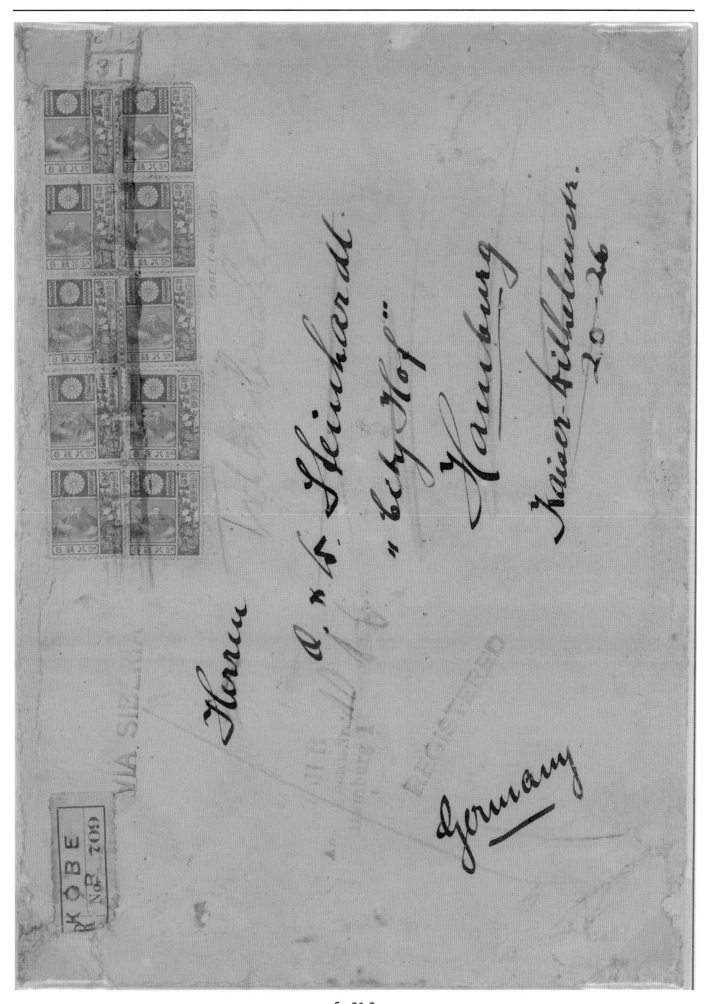

fig.31-3
1929 シリーズ（富士鹿・改色）DIE I のマルチプル 8 銭（10 枚ブロック）外信書留書状 KOBE → ドイツ 20 12 1931
The largest multiple of 1929-1937 issue 8 Sen, DIE I , Registered letter from KOBE to Germany, 20 12 (1931)

２０銭

単線１１目打バラエティ

　富士鹿切手のスター的存在が、この単線１１目打です。なぜなら原則として櫛型目打 13 x 13 ½ で穿孔される富士鹿切手における唯一の例外だからです。

　耳紙や銘版のついた切手が発見されていないため、製造時期の決定打はありませんが、同時代の通常切手である田沢切手における単線11目打の使用時期から推定すると、1924 年つまり第２期製造分の印刷後に一部のシートが単線目打穿孔されたと推定されています。

　使用局は神戸と聖護院（京都）が多く、使用時期は、1925 年から 1928 年に分布しています。

　このバラエティは一時期は珍重されていましたが、数多く掘り出されてきて、既に 70-90 枚発見されていると推測しています。一方でペアを越えるマルチプルの発見数はまだ少なく、５枚ストリップ２点、３枚ストリップ１点、田型２点のみを確認しています。

　使用例は３点発見されており、うち２通が６銭田沢切手との混貼り外信書状（図９）で、残りの１通が混貼り外信書状です。但し見逃す可能性のあるバラエティですので、今後もまだまだ発見される可能性は高いと思います。

20 Sen

Line Perf.11

The most attractive item of the 1922-1937 issue is line perf.11 of the 20 Sen of the 1922 issue. It is the only exception of the whole series, which generally has the comb perf.13 x 13 ½.

No marginal stamps has been found, and we have no decisive element to show the producing period. Howver, line perf. 11 of the Tazawa issue tells us that it was produced in 1924, during the second production period between 1924 and 1925.

They were mainly used at Kobe and Shogoin (Kyoto) post office between 1925 and 1928.

This variety was highly valued before, but as much as 70-90 examples have been found in the postwar period. However, multiple more than pairs are still scarce, and only two strips of five, a strip of three and two blocks of four are known.

Three covers are known, two of which are international letters franked with a 6 Sen Tazawa stamp (fig.9), and the rest one is an international letter with other stamps. I think there are big possibility to find new examples in the future.

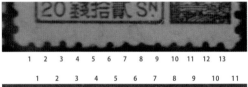

fig.33 単線 11 目打バラエティ 未使用　上は比較の為の P.13
L 11 only one known unused,
the above illustration is　P.13 for reference.

fig.34 単線 11 目打バラエティ 最大ブロック KOBE2
The largest multiple of L.11, block of four KOBE2

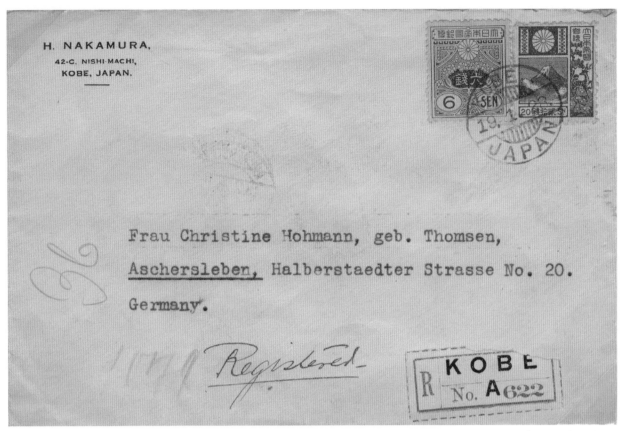

fig.34 単線 11 目打バラエティ、田沢 6 銭との混貼り外信書状。KOBE to Germany 19 1 28 カバーは現存 2 点
L.11 franking KOBE to Germany 19 1 28

fig.35 単線 11 目打バラエティ、現存 2 点の最大マルチプル（5 枚ストリップ）KOBE 3 7 25 (Taisho 15) 同時使用
L.11 The largest strips of 5, KOBE 3 7 25

横穴のある目打バラエティ

単線11目打バラエティと順番が前後しますが、1922年の旧版富士鹿第1期の中には、シートの左右マージンに、横穴のある目打バラエティが存在します。

前ページで記載した通り、大半の富士鹿切手の櫛型目打穿孔の結果では、左右の耳紙に穴が開けられてはいません。しかし1902年に導入され、当時の現行切手である菊切手や田沢切手に多くみられる櫛型目打は、そもそも左右の耳紙に穴が一つ飛び抜けて穿孔される形状をしていました。

櫛型目打の穿孔に使われる針からこの穴がなくなった時期は、複数存在すると推定される目打穿孔機により異なります。

1920年代前半に製造されたと推定される切手でも既に穴がないものもありますが、1922年シリーズ初期の製造時期には最低一台の目打穿孔機に横穴がまだ残っていたようで、私は20銭にしか確認していませんが、(fig.36) 4銭にも存在することが知られています。

Perforation Variety with Side Holes

This variety appeared before the variety of L.11. The first production of the 1922 issue includes a perforation variety of sheets with side holes on both sides.

Major comb perforations of the 1922-1937 issue didn't have any holes on both sides of the margin. However, the comb perforation, which was introduced in 1902 and used mainly for the Chrysanthemum or the Tazawa issues, originally had one additional hole on both sides.

The time when these side holes disappeared varies with the perforation machines, which were assumed to exist a few.

Some sheets without side holes were assumed to be produced in early 1920's, but sheets of the 1922 issue in its early period also have holes like fig.36, I have only 20 Sen examples, but 4 Sen is also known with this perf. variety.

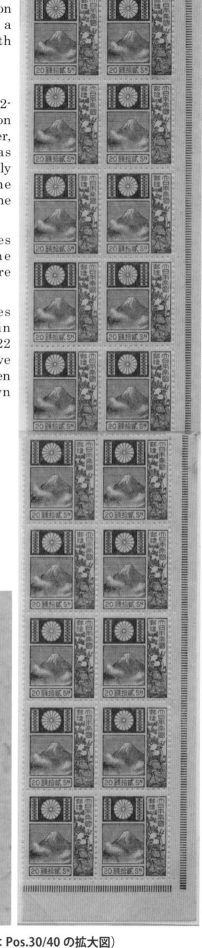

fig.36 横穴のある目打バラエティ（上は Pos.30/40 の拡大図）
Perf. variety with a side hall, enlarged illustration of pos.30/40

エラー、フリーク

Error and, e.t.c.

Error and, e.t.c.

富士鹿切手の製造不具合はこれまでまとめて発表されたことがなく概要が分かっていませんでした。人気がないシリーズながらいくつか作品集を発表した方もいらっしゃいましたが、それらの中にエラーと呼べるマテリアルが入っていることもありませんでした。

ただ、オークションや即売には時々この手のマテリアルがポツンと登場することがあります。私が富士鹿切手の収集を開始してから7年ですが、その間に見つけたエラーやフリークは全て入手しましたので、最後に紹介したいと思います。

No report on erros and, e.t.c. exist for the 1922-1937 issue before. Although the series was unpopular, a few philatelists published their collections, but such collections didn't include errors and, e.t.c.

I bought every error stamps of this issue at public auctions or buy-it-now sale in this seven years since I started collecting this series. I'd like to show all of them at once.

Printed on pre-folded papers
ペーパーフォールド

Perforated on pre-folded paper
ペーパーフォールド目打穿孔

Horizontal Shift of the Perf.
目打穿孔左右ズレ

Double Print or Double Transfer 二重印刷

This error stamp was a result of printing with another piece of paper on. most of the design was printed on the other piece of paper, and that part of the sheet remained white. 別用紙片混入

まとめ

富士鹿・風景切手は、私が 2007 年に切手収集を再開した後に初めて本格的に伝統収集の対象として取り組んだ日本切手です。2013 年 11 月に収集を開始しましたので、もう 7 年になります。

デザインが好きなことが一番ですが、この分野に特化した収集家が 1 人も居ないブルーオーシャンだったことから、国際展を狙ってみようと思い収集を開始しました。

ここに至るまでに、横矢仁さん（広島、故人）に最初の手ほどきを受けた他、槇原晃二さん（広島）、杉山幸比古さん（東京）、山口充さん（千葉）、林国博さん（長野）を始めとする多くの方にコレクションを見せていただき、解説していただきました。

私の現在のコレクションは上述した方々を始めとする多くのフィラテリストの収集と研究を元にした知識の現時点での総括の上に成立するものです。

7 年前に比べると、収集される方も少しずつ増えてきて、珍しいマテリアルはオークションでも高く落札されることも増えてきましたが、それでも前後のシリーズに比べてまだまだ集めやすいと思います。

是非、本記事をご熟読いただき、製造面に軸足を置いたアプローチで国際展を目指す方が現れることを希望します。

Conclusion

The 1922-1937 issue is the first Japanese stamp series which I've tried to compose traditionally after my come back to philately in 2007. I started it in November, 2013, so 7 years has passed.

I like this series just because I like the design, but I also decided to aim international competitions with this series just because it was a BLUE OCEAN without any rivals for the seris.

I was first taught by Mr. YOKOYA how to collect this issue, and Mr. MAKIHARA, Mr. SUGIYAMA, Mr. YAMAGUCHI, Mr. HAYASHI also showed me their collections and taught me their materials.

My collection stands on the knowledge of the above mentioned philatelists.

The number of philatelists having interests on the issue is gradually increased, and rare materials went higher than expected at auction sale some times, however, it is still reasonable in comparison with the issues before and after the 1922-1937 issue.

The author hopes that more and more philatelists have interests on the issue and aim international exhibitions traditionally.

参考文献 Ref.

富士鹿・風景切手（吉田 敬 2017, スタンペディア）、富士鹿・風景（1993, JPS）

Japan Definitive Issues 1922-1937, Takashi YOSHIDA, 2017, Stampedia　　Fuji shika and Fukei series, 1993, JPS

The Author：YOSHIDA Takashi

Collecting：World General up to 2005

Awards：

　Kingdom of Prussia ‐LG (WSC ISRAEL 2018)

　Classic Switzerland ‐LV (PHILAKOREA 2014)

Membership & Qualification and etc.：

　STAMPEDIA PROJECT FOUNDER since 2009

　Society for Promoting Philately (Japan) Representitive Director

　Club de Monte Carlo, member since 2014

　Consilium Philateliae Helveticae, AEP, AIJP

ゴーティエ・フレーレ社ラベルの使用例

Private Ship Letter Label of Gauthier Frères et Cie

正田幸弘 SHODA Yukihiro

1851 年、英国政府からの補助金を受けている Royal Mail Steam Packet Company は、サウサンプトンからリオデジャネイロへの月 1 便の定期航路を開設しました。英国政府は月 1 便の郵便船を 1808 年にファルマスからリオデジャネイロへ就航開始していますが、民間契約船による輸送に方針変更した訳です。

一方フランスでは 1853 年にマルセイユ発の Compagnie de Navigation Mixte が、1856 年にはルアーブル発の Compagnie Franco-Américaine がブラジル航路を短期間ながら運航を開始しました。

特に、最後の会社は船図案のラベル (fig.1) の使用例がありますので、郵便史収集家以外の船切手やローカル切手の収集家にも注目の対象です。(ref.1)

In 1851, Royal Mail Steam Packet Company, which the British government subsidized, started a monthly liner service between Southampton and Rio de Janeiro. it replaced a monthly ship mail between Falmouth and Rio de Janeiro which British government started itself in 1808,

In France, on the other hand, Compagnie de Navigation Mixte opened the Brazil route from Marseille in 1853, and Compagnie Franco-Américaine, from Le Havre in 1856, even for a short period.

The latter one's private ship letter labels, especially, attract not only postal history collectors but ship stamp collectors or local stamp collectors (ref.1).

fig.1

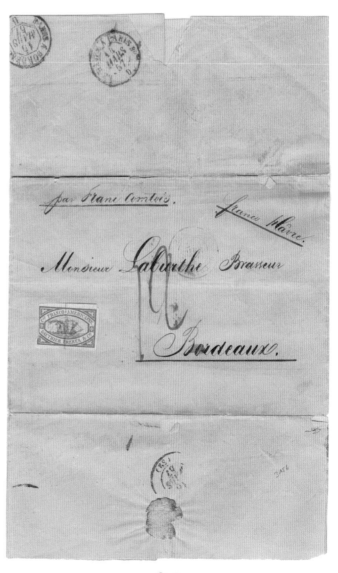

fig.2

フランコ・アメリカーナ郵船

リヨンに本社を持つフランス系アメリカ人による Gauthier Frères et Cie は、1856 年 パ リ で Companie Franco-Américaine de Frères Gauthier e Cie を設立し、貨物と旅客の輸送とフランス政府から郵便物の輸送許可を取得しました。8 つの船によって、ニューヨークへ 1856 年 2 月から翌年 1 月まで 12 航海、リオデジャネイロへは 1856 年 2 月から 9 航海【tbl.1】、さらにアンティル諸島とニューオーリンズへ 1856 年に 3 航海、ハバナとプエルトリコへ 1857 年 4 月から翌年 3 月に 11 航海、実施しました。

The Mail Ship, Franco-Américaine

In 1856, Gauthier Frères et Cie, a French American joint venture established in Lyon, founded the Companie Franco-Américaine de Frères Gauthier e Cie in Paris. It got the French government transport permission of goods and passengers. They operated the service with eight ships: twelve sailings to New York from February in 1856 to January the next year, nine to Rio de Janeiro from February in 1856 [tbl.1], three to the Antilles and New Orleans in 1856, and eleven to Habana and Puerto Rico from April in 1857 to March the next year.

Havre départ	Paquebot	Ténériffe	Rio départ	Lisbonne départ	Havre arrivée
26/02/56	Cadix	10/03/56	10/04/56	11/05/56	16/05/56
05/04/56	Lyonnais	17/04/56	18/05/56	14/06/56	18/06/56
08/05/56	Franc-Comtois	18/05/56	18/06/56	17/07/56	23/07/56
06/06/56	Cadix		18/07/56	14/08/56	20/08/56
02/07/56	Lyonnais		10/08/56	15/09/56	20/09/56
02/08/56	Franc-Comtois		10/09/56	17/10/56	22/10/56
18/09/56	Cadix		31/10/56	25/11/56	30/11/56
23/10/56	Barcelone		05/12/56	09/01/56	15/01/57
03/12/56	Franc-Comtois		05/02/56	07/03/56	14/03/57

Tableau de marche des navires de la "Compagnie Franco-Américaine"/

tbl.1　参考文献 2 ref.2

サービスの全期間中、同社はフランス政府の援助なしに独自に運営され、補助金付きの英国企業と競争しました。海運会社の収入源は 3 つ（貨物・旅客・郵便）ですが、旅客輸送と郵便物はスピードと時間厳守が不可欠です。同社は大西洋横断に 35 日かかり、英国ロイヤルメール社の 25 日よりも時間がかかりました。

1856 年 11 月 1 日、193 人の乗客乗員を載せてニューヨークを出港した Lyonnaise 号が翌日、米国の帆船 Adriatic 号と衝突し、120 名死亡という海難事故を起こし、会社は経営難に陥ります。（ref.2）

In all the service period, they managed by themselves without the French governmental support and competed with the subsidized British companies. A shipping company has three income sources: goods, passengers, and mails, and the transport of passengers and mails requires speed and strict punctuality. Franco-Américaine needed 35 days to sail over the Atlantic Ocean; it was more than Royal Mail of Britain, which took 25 days.

Then, their ship Lyonnaise, which left New York with 193 passengers and staff on November 1, 1856, collided with the U.S. sailing vessel Adriatic the next day. It was a grave maritime accident that took 120 lives, and the company had fallen into financial difficulties (ref.2).

郵便物の実例

Usages of Franco-Américaine

A. 1857 年ボルドー宛　赤ラベル貼（第9航海）

A. To Bordeaux in 1857, with a red label, 9th voyage

fig. 2　筆者蔵。Eugene Klein, Julius Steindler, Renato Mondolfo, Bernald Berkinshaw-Smith, Jan Berg 旧蔵。

ボルドーの Monsieur Laburthe Brasseur 宛。カバー表面右上に franco Havre、左上に par Franc Comtois の書き込みがあります。中央にルアーブル 1857 年 3 月 13 日の赤の二重丸入国印つき。tbl.1 にある 2 月 5 日リオ出港の Franc-Comtois 号と一致します。

ブラジルからフランス宛ての手紙料金は 1849 年 8 月 1 日から 7.5g 当たり 60 サンチームだったので、この手紙は 7.5 〜 15g の二倍重量で、"12" から 12 デシームを受取人から徴収したことがわかります。

リオの MI.I.Seiler が 1857 年 2 月 4 日に書いた手紙で、裏面に 3 個のハンコ Le Havre a Paris 14 Mars 57、Paris a Bordeaux 15 Mars 57 と Bordeaux 16 Mars 57 から，入国後も順調に鉄道で逓送されたことがわかります。

fig.2 The author's collection. Ex-collection of Eugene Klein, Julius Steindler, Renato Mondolfo, Bernald Berkinshaw-Smith, and Jan Berg.

Mail to Monsieur Laburthe Brasseur in Bordeaux. It has the handwriting of "franco Havre" at the right top of the cover and "par Franc Comtois" at the left top. In the middle, a red double-circle entry mark of Le Havre dated March 13, 1857. The date is the same as Franc-Comtois' schedule in tbl.1, which departed Rio de Janeiro on February 5.

The postage of a letter from Brazil to France was 60 centimes up to 7.5g from August 1, 1849. So, the letter must have been a double weight between 7.5 and 15g. And the writing "12" shows the addressee paid 12 décimes at receiving.

It was a letter written by MI.I.Seiler in Rio de Janeiro on February 4, 1857, and it has three seals: "Le Havre a Paris 14 Mars 57," "Paris a Bordeaux 15 Mars 57," and "Bordeaux 16 Mars 57," which show that it was delivered safely in France by railway.

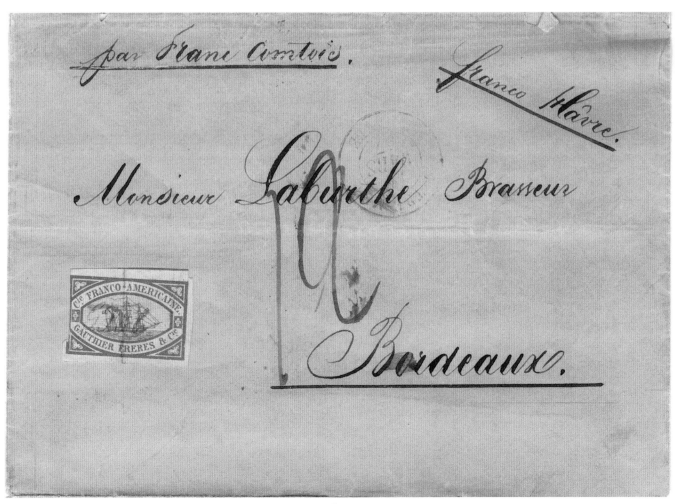

fig.2

fig. 3 筆者蔵。Reinhold Koester, Bernard Berkinshaw-Smith 旧蔵。

A のカバーの4カ月前に同じ船で運ばれたカバーです。パリの Rothschild Brothers 宛で、宛名の下に per Franc Comptois と書かれています。右上のルアーブル入国印も 1856 年 10 月 22 日でリオ9月10日出港と一致します。

"6" から重量 7.5g 以下の手紙で、6 デシームを受取人から徴収したことがわかります。さらに青色の船会社のバイア代理店のハンコもあります【fig.4】。

裏面には Le Havre a Paris 22 Oct 56 の鉄郵印があります。この手紙には、バイア 1856 年 9 月 15 日の書き込みがあり、手紙の内容も通貨交換やダイヤモンドの出荷が書かれ、通常の商業通信であることが、London Philatelist の論文でも強調されています。（ref.3）

fig.3 The author's collection. Ex-collection of Reinhold Koester and Bernard Berkinshaw-Smith.

It is a cover delivered by the same ship of CASE A, four months before that. It was to Rothschild Brothers in Paris, and it has the handwriting "per Franc Comptois" under the address. Le Havre entry mark at the right top was on October 22, 1856, and it corresponded with the schedule of departure from Rio de Janeiro, on September 10.

The writing "6" means it was a letter weighing 7.5g or less, and the addressee paid 6 décimes to receive it. It also has a blue mark (only used at Bahia) of the shipping agency Bahia [fig.4].

It has a railway postmark of "Le Havre a Paris October 22 56" on the back. The letter has the handwriting of "Bahia, September 15, 1856," and the content includes the story of currency exchange and shipping of diamond. Article on "London Philatelist" also emphasized that it was an ordinary commercial correspondence (ref.3).

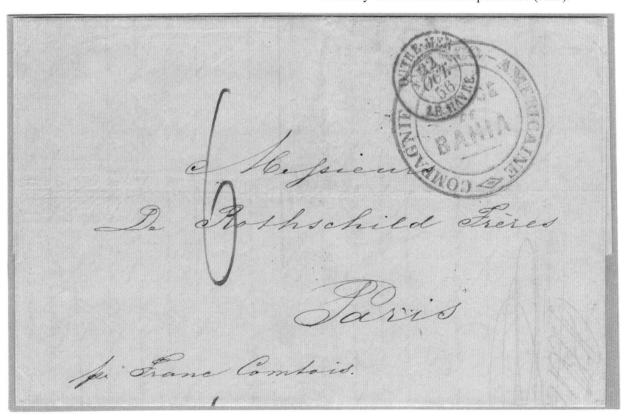

fig.3

"Compagnie Franco-Américaine"

Cachet administratif/Carimbo
administrativo

Vignettes /Vinhetas

fig.4　参考文献2 ref.2

Auction Preview September 2020 - Corinphila Auctions より

"Auction Preview September 2020" of Corinphila Auctions

fig.5 Ritchie Bodily, Julius Steindler, Renato Mondolfo, Gregory Frantz 旧蔵

fig.5 Ex-collection of Ritchie Bodily, Julius Steindler, Renato Mondolfo and Gregory Frantz

上部に Voie de Havre par_le_Franc-C__ の書き込みや Outre Mer 13 Mars 57 の赤印、裏面の Le Havre a Paris 14 Mars 57 と Paris a Bordeaux 15 Mars 57 の印から、A と同じ船と逓送経路と推定できますが、どこからの発信かはわかりません。また "12" から、二倍重量便の 12 デシーム料金です。

It has the writing of "Voie de Havre par_le_Franc-C__" at the top, a red postmark of "Outre Mer 13 Mars 57," and the postmarks on the back, which were "Le Havre a Paris 14 Mars 57" and "Paris a Bordeaux 15 Mars 57." All those shows it was delivered on the same ship and route of CASE A, but we haven't found the place of dispatch. And the writing "12" shows the postage was 12 décimes, for the double weight.

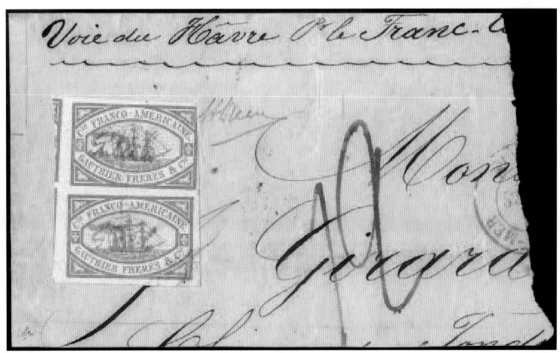

fig.5

D.1856 又は 1857 年リスボン宛 青ラベル貼 フロントカバー（第 3,6,9 航海のいずれか）

D. To Porto in Portugal, in 1856 or 1857, with a blue label. Front cover, either 3rd, 4th, or 9th voyage

fig. 6 ロペス氏所蔵品 Charles Nissen, Philipp von Ferrary, Eugene Klein, Julius Steindler, Renato Mondolfo, Bernald Berkinshaw-Smith 旧蔵。

宛先が Duarte Irmãos & Co, Lisbon、さらに per Franc Comtois の書き込み。"440" は郵便料金です。ブラジルからポルトガルへの海上郵便料金は、リスボンに支店を持つ Royal Mail Steam Packet Company のような大手とは異なり、中小の郵船会社の運ぶ郵便物には PT (Paquetes Transatlânticos nâo Subsidiados) Packet Rate が適用され、1 オンス（28.35g）の 8 分の 7 の重量の手紙が 440 レイスでした。

したがって約 21 〜 25g くらいの重さで、受取人から 440 レイス徴収したことがわかります。残念ながら郵便印が押された部分がないのですが、船名からリスボン到着は 1856 年 7 月 16 日、10 月 16 日、1857 年 3 月 7 日のいずれかと推定できます。（ref.4）

fig. 6 The collection of Mr. Lopes. Ex-collection of Charles Nissen, Philipp von Ferrary, Eugene Klein, Julius Steindler, Renato Mondolfo, and Bernald Berkinshaw-Smith.

The addressee was Duarte Irmãos & Co, Lisbon, and it has the writing of "per Franc Comtois." "440" is the postage. The sea mail postage from Brazil to Portugal applied PT (Paquetes Transatlânticos nâo Subsidiados) Packet Rate for the mails delivered by small and medium-sized mail companies. But it was not applied for the mails delivered by a large company such as Royal Mail Steam Packet Company, which had a branch in Lisbon. And 440 reis was the postage for a letter of the seven-eighths weight of 1 oz (28.35g).

Therefore, the letter weighted between about 21g and 25g, and the addressee paid 440 reis to receive it. Unfortunately, it doesn't include the postmark part, but by considering the ship name, the presumed arrival date in Lisbon is July 16 or October 16, 1856, or March 7, 1857 (ref.4). This is the only known with a blue label.

fig.6

fig.7 Franc Comptois の書き込みがあり、Ｂと同じバイア代理店の二重丸印つき。リスボンで３月７日の"P.TRANSATLANTICO"印、ポルトで 1857 年３月９日の着印。A,B とおなじ第９航海で運ばれています。

1853 年 Compagnie de Navigation Mixte の運行開始で PT 郵便料金が設定され、１オンスの８分の２の重量の手紙が 160 レイスでした。約５〜７g の重さで、受取人から 160 レイス徴収されたことが "160" の表示でわかります。

fig.7 It has the writing of "Franc Comptois," with the same Bahia agency's double-circle mark of CASE B. It has Lisbon "P.TRANSATLANTICO" mark of March 7 and Porto arrival mark of March 9, 1857. The 9th sailing delivered it with the letters of CASE A and B.

In 1853, when Compagnie de Navigation Mixte started service, PT postage was established. And the postage for a letter of the two-eighths weight of 1oz was 160 reis. The writing of "160" shows it weighed about 5 to 7g, and the addressee paid 160 reis at receiving.

fig.7

ゴーティエ・フレーレ社製ラベルの使用状況

こぎれいな船図案のラベルは上記 3 通の使用例以外では、赤ラベルの単片 4 枚が知られています。(ref.4)

中央の図案は Barcelone 号で、上部に "CIE FRANCO-AMERICAINE", 下部に "GAUTHIER FRERES & CIE" と書かれていますが、額面数字等はありません。ラベルは平版印刷で無目打ち。印刷者やシート構成は不明。未使用は存在せず、ラベルの抹消は "G.F.& Co." の文字で青か黒で行われています。

3 通の使用例は赤の単片、赤の縦ペア、青の単片でその使用目的は不明です。

赤ラベルがブラジル路線用、青ラベルがアメリカ路線用との意見もありましたが（ref.5）、D の使用例と矛盾します。所有者のロペス氏は赤ラベルがフランス行きに、青ラベルがポルトガル行に使われたとの意見です。

氏は本稿での ABCD4 例を含む 11 例の郵便物を検討されていますが、うち 2 通はフランス切手貼付ですが、それ以外はスタンプレスで郵便料金は受取人払いです。

tbl.1 にある 9 航海のうち第 2、第 5 航海の Lyonnais 号以外はすべて存在します（ref.6）。私製ラベルが貼られた 3 例はすべて Franc-Comtois 号ですが、第 6 航海のバイア発（B 例）にはラベルはありません。もしかすると D 例を含め 3 通とも最終第 9 航海なのかもしれません。

ここで一つの仮説を提案したいと思います。「ラベルはブラジル発最終航海の Franc-Comtois 号のリオ積み込みの郵便物に貼られ、青ラベルはリスボン揚陸、赤ラベルはルアーブル揚陸の区別がある。」

この仮説の妥当性の検証には、ラベル貼付のカバー出現が望まれますが、ラベルの無いスタンプレスカバーの収集によっても、真相に近づける可能性があると思います。

郵趣史上の位置づけ

この魅力的な小紙片はウィリアムズ兄弟の『シンデレラスタンプス』（1970）にも巻頭で取り上げられていますし、フェルドマンから出た珍品切手本でも詳細に扱われています。

今日では収集家目当ての作り物だという人はいないでしょう。しかし、ニッセンから購入を持ち掛けられたメルビルは即断できず、フェラリ収集に入ったと『ファントムフィラテリー』（1923）に書いています。【fig.8】

1925 年のフェラリ・セールではロット中に紛れていたので一時行方不明となったそうです。1925 年に購入した Eugene Klein が 1943 年にこうした経緯を公表しました。もともとはアメリカ私設郵便の専門家 Henry Needham の

Use of the Private Ship Letter Labels of Gauthier Frères et Cie

The known usages of the private ship letter labels other than the above three are four single-use of red labels (ref.4).

The label has the drawing of the ship, Barcelone in the middle, "CIE FRANCO-AMERICAINE" at the top, and "GAUTHIER FRERES & CIE" at the bottom. But it doesn't have the description of face value. The labels were printed by the lithographic press, without perforation. The printing factory and sheet composition are unknown. Any unused label does not remain, and the blue or black handwriting of "G. F. & Co." canceled the labels.

The above three usages are a red single, a red vertical pair, and a blue single. The purpose is unclear.

Some philatelists said the red label was for the Brazil route and the blue for the U.S. one (ref.5), but that doesn't correspond to the CASE D. The owner of the label of CASE D, Mr. Lopez, has an opinion that they used the red labels for the ship to France and the blue ones to Portugal.

He has studied the mails of 11 cases, including the other four cases of this article. All mails, but two with French stamps, were without label, and the addressee paid the postage.

In the nine sailings shown in tbl.1, the usages of all sailings, but Lyonnais of the 2nd and 5th sailings, have been found (ref.6). Franc-Comtois delivered all the three mails with private ship letter labels, but the 6th sailing from Bahia (CASE B) had none of them. So, all the three letters, including CASE D, could be delivered by the last sailing, the 9th one.

Now I propose a hypothesis. "**They used the private ship letter labels for the mails that the last sailing of Franc-Comtois took on board and delivered from Rio de Janeiro, Brazil. The blue labels were for the mail to Lisbon and the red to Le Havre**."

We need more covers with private ship letter labels to verify this hypothesis, but the covers without them will also help to find the truth.

Private Ship Letter Labels in a History of Philately

These attractive labels appear at the beginning of "Cinderella Stamps," written by William brothers (1970), and Feldman's book of rare stamps describes the details.

Now no one says they are forgeries for collectors. But Melville confessed in his book "Phantom Philately" (1923) that when Nissen had proposed him buying one, he couldn't have decided it immediately and it had gone to Ferrary [fig.8].

At the Ferrary sale in 1925, it was mixed in one lot and missing for a while. Eugene Klein, who bought it in 1925, told this story in 1943. Initially, a stamp dealer in New York, Eugene Coastale, found these labels in the collection of a U.S. private mail specialist, Henry Needham.

コレクションからニューヨークの切手商 Eugene Coastale が発見したものです。

　メルビルの時代には船会社の運行表も未発見だった訳で、確信が持てなかったのでしょう。というのも 1877 年にスコット商会が The United States Locals and Their History を発行した際、著者の Charles Henry Coster は「いくつかの切手は想像から作られ、投機的利益用である」と説明したうえで、わざわざ図示したうえで、こうした名称の船会社自体が存在しないのだから、切手は架空品であると断じたそうです。（ref.7）【fig.9】

　私もかつてこのラベルの記事を書いたことがあります（ref.8）。最初はエラー等の希少切手からスタートした連載が、だんだんと郵便史上の特殊性へと興味の対象が変わってきた経験があります。今日、郵便史収集が益々盛んになっています。スタンプレスカバーは切手つきのカバーに比べて見た目の派手さがありませんが、収集対象としては色々な可能性があるもんだと思っています。

In the age of Melville, since the sailing schedule hadn't been discovered yet, he was not sure. I say that because in 1877, when Scott published The United States Locals and Their History, the author Charles Henry Coster said, "some stamps were invented by imagination for speculative profit." He especially illustrated the figure and condemned the stamp was a forgery because the steamship company of that name didn't exist (ref.7) [fig.9].

I've written an article for this label before (ref.8). I started a series of articles with rare stamps such as error and was changing my interest in the particularity of postal history. Postal history collection becomes popular nowadays. Stampless covers are not so good looking like covers with stamps, but they have the various possibility as a collection.

Gauthier Frères et Cie. I have dealt with this local in "Stamps of the Steamship companies" (1915) and in the STAMP LOVER (VI, 31, 199), for it is difficult to accept here the statement of Coster that the Company never existed. On page 2 of my steamship booklet I illustrate a copy used on cover, which was offered to me when a boy, and although I wanted to keep it the price was beyond me, and it went to Ferrary at three times the figure it was offered to me. It was as convincing a cover as one could wish for, and I am of opinion that Coster's denunciation was based upon too little inquiry. Most of the copies of the stamp seen nowadays are bad, but are probably forgeries of a stamp—of extreme rarity—which actually existed prior to 1864.

fig.8　Phantom Philately, P.103　Melville (1923)

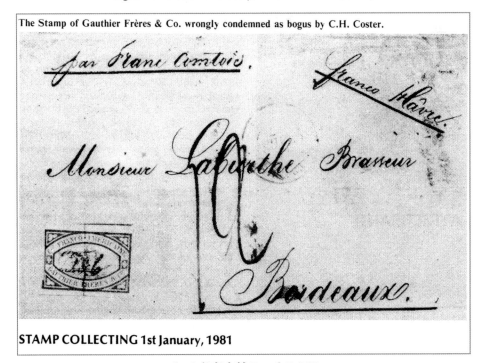

The Stamp of Gauthier Frères & Co. wrongly condemned as bogus by C.H. Coster.

STAMP COLLECTING 1st January, 1981

fig.9 参考文献 7　ref .7　P47

　友人のロペス氏に本記事の改善も兼ねて記事原稿を見せたところ、所蔵のデータ 5 例の掲載許可をいただきました。すべてバイア発信です。

When I show this article to Mr. K. Lopes to improve the contents, images of 5 covers of his collection were sent with a pemission to list in this article. All sent from Bahia.

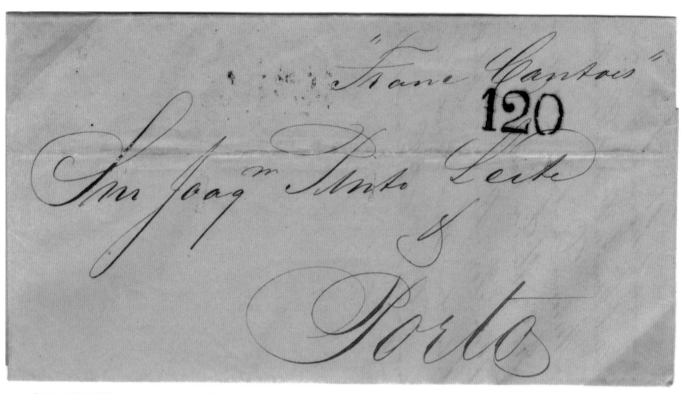

F.　ポルト宛 第 3 航海 Franc-Comtois 号 バイア 1856 年 6 月 21 日、リスボン 7 月 16 日、ポルト 7 月 19 日 120 レイス（3.5-5g 位）支払
　　BAHIA,1856/ 6/21 → Franc-Comtois → Lisbon, 7/16 → Porto, 7/19 The third voayage, 120 R.

G.　ロンドン宛 第 3 航海 Franc-Comtois 号 バイア 1856 年 6 月 20 日、ル・アーブル 7 月 23 日、ロンドン 7 月 24 日 2 シリング 3 ペンス支払
　　BAHIA, 1856/6/20 → Franc-Comtois → Le Havre, 7/23 → London, 7/24 The third voayage, 2 sh. 3 d

H. パリ宛 第 7 航海 Cadix 号 バイア 1856 年 11 月 2 日、ル・アーブル 11 月 30 日、パリ着 12 デシーム（15-30g）支払
BAHIA, 1856/11/2 → Cadix → Le Havre, 11/30 → Paris The seventh voayage, 12 desimes

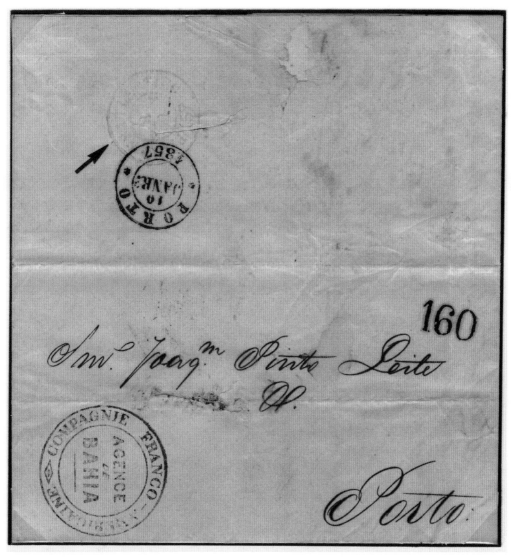

I. ポルト宛 第 8 航海 Barcelone 号 バイア 1856 年 11 月 26 日、リスボン 1857 年 1 月 8 日、ポルト 1 月 11 日
BAHIA, 1856/11/26 → Barcelone → Lisbon, 1/8 → Porto,1/11 The eighth voayage

J. パリ宛 第 8 航海 Barcelone 号 バイア 1856 年 11 月 27 日、ル・アーブル 1857 年 1 月 15 日、パリ着 6 デシーム（15g 以下）支払
BAHIA, 1856/11/27 → Barcelone → Le Havre, 1/15 → Paris The eighth voayage, 6 Decimes

参考文献 / References

（ref.1） 正田幸弘『ブラジル郵便史概説』(2010)　pp23-32 契約郵便船の時代へ、pp44-46 各種郵便船

（ref.2） K.Lopes, Caractéristiques des correspondances du Brésil pour la France au 19ème siècle（2003）pp75-80 Les lignes françaises de navires à vapeur avant 1860【tbl.1】【fig.4】

（ref.3） B.Berkinshaw-Smith, "The Carriage of Mail Between France and the Antilles and Latin America to 1880" London Philatelist Vol.102 (May1993) pp139-150

（ref.4） L.N.Williams, Encyclopaedia of Rare and Famous Stamps Vol.2 The Biographies (1997) pp78-82 Gauthier Frères et Cie

（ref.5） S.Ringstron and H.Tester, The Private Ship Letter Stamps of the World Part2(1983) pp149-152 Gauthier Frères et Cie

（ref.6） K.W.Lopes," O Serviço da Companhia Franco-Americana no Brasil, de 1856 a 1857" (2007) 11pp

（ref.7） L.N.& M.Williams, "Philately 100 years Ago"Part I　Stamp Collecting　1st Jan.1981 p46-47,49,51,53,55

（ref.8） 正田幸弘『新・紙の宝石』（2002）New Jewels of Paper, (Yukihiro SHODA, 2002) p145 Gauthier Frères et Cie

The Author： SHODA Yukihiro

Collecting：Brazil 19th Century

Awards：Postal History of Brazil-LG

　　　　　　(China 2009, Bangkok 2010, Thailand 2016)

Membership & Qualification and etc.：

　FIP Accredited Jury (PH since 2001, LI since 2011)

　American Philatelic Society since 1980

　Royal Philatelic Society, London since 1983

　The Collecters Club since 1988

在横浜フランス郵便局
横浜外国人居留地発着郵便物
French Post Office in Yokohama
The Mail from/to Foreign Settlement in Yokohama
小林 彰 KOBAYASHI Akira

はじめに

1853年7月8日、ペリー提督率いる4隻の黒船が浦賀沖に来航し、日本政府（徳川幕府）に国交を求めました。ペリーは開国を促すフィルモア大統領親書を幕府に手渡しましたが、彼らは返答まで1年を要求したため、ペリーは1年後に再来航すると応じました。

そして1854年2月13日、ペリーは琉球を経由して浦賀沖に再び姿を現しました。幕府との取り決めでは1年間の猶予を与える筈でしたが、半年で決断を迫ったのです。約1か月にわたる協議の末、幕府はアメリカの開国要求を受け入れ、全12条におよぶ日米和親条約が締結されました。徳川家光以来200年以上続いた鎖国の終焉です。その後、列強は相次いで和親条約を締結し、外交事務と居留民保護のため公使館や領事館を開設しました。

さらに1858年になると、幕府は5ヵ国、アメリカ、イギリス、ロシア、オランダ、フランスと相次いで、修好通商条約を結ぶに至りました。ポルトガル、イタリア、デンマーク、スペイン、スウェーデン、ノルウェー、オーストリア・ハンガリーほかも5か国に遅れること数年、同様の条約に調印しました。

Introduction

On July 8, 1853, the four steamships called "black ships" under Commodore Matthew Calbraith Perry visited the Uraga shore to ask diplomatic relations to the Japanese government (Tokugawa Shogunate). Perry handed the government the personal letter from President Fillmore. The Japanese government asked for a year to answer it, and Perry responded to come back then.

On February 13, 1854, Perry revisited the Uraga shore via Ryukyu. Against the agreement of one-year postponement with the government, he pressed for an answer in half a year. They deliberated for about one and half a months, and the government agreed to open the country. They concluded the Japan-US Convention of Peace and Amity consisting of twelve articles. It was the end of the national isolation, which had lasted more than 200 years from the era of Iemitsu Tokunaga.

Then, other great powers also concluded Convention of Peace and Amity with Japan, and they established legations and consulates to handle the paperwork of diplomatic relations and protect their national residents.

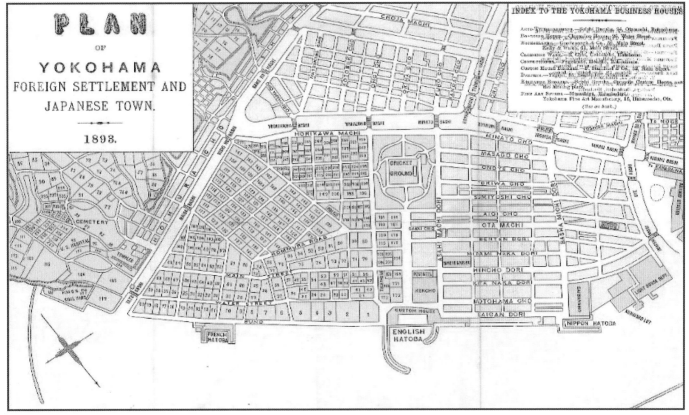

fig.1 フランス人測量技師クリペー制作居留地地割 (ケンソン氏所蔵) Settlement Allocation Map Made by a French Surveyor, Clipet

この修好通商条約により、横浜、長崎、函館、兵庫と新潟の５港で自由貿易が認められました。自由と言っても条約締結国の国民に借地と住居、営業が許可されるのは、開港場内の一定地域に制限されていました。この地域を「居留地」といいます。借地権や建物の所有権を含む外国人の居住権を「居留」と呼んだことから、この名がつきました。横浜の居留地は現在の山下町と山手の丘一帯にあり [fig.1]、ここには公使館、領事館、商館が建ち並び、また瀟洒な異人館が軒を連ねていました [fig.2]。

欧米から大勢の外交官、御雇外国人、聖職者、商人等が開港地にやって来て、居留地内に落ち着きました。当時、日本ではまだ近代郵便制度が未成立で居留民は本国との通信には領事館を通じるしか手立てがありませんでした。これを領事館郵便といいます。領事館員が本務の片手間に郵便事務を取り扱った程度のものから、後年の英仏横浜局のように独立した局舎と専従職員を有する組織的なものまで、その進化の程度には大きな差がありました。

本稿では、在横浜フランス郵便局の開局から閉局までの所在地の変遷と局長の任にあったアンリ・デグロン関係に言及し、また、居留地発着郵便物を紹介します。

In 1858, the government had an amity treaty with five countries, such as the U.S., Great Britain, Russia, Holland, and France. Portugal, Italy, Denmark, Spain, Sweden, Norway, Austria-Hungry, and other countries also concluded a similar treaty with Japan later.

These amity treaties permitted a free trade at the five ports; Yokohama, Nagasaki, Hakodate, Hyogo, and Niigata. However, the area where the people of the contracting countries were free to lend, resident, and do business was limited in a specific part of a treaty port. The area was called "Kyoryuchi (Settlement)." The name was derived from "Kyoryu," meaning the foreigners' right of residence, including leasehold or ownership of a building. The settlement in Yokohama was on the whole hill around now Yamashita-Cho and Yamate [fig.1], and there were legations, consulates, trading houses, and elegant Western-style residences there [fig.2].

A lot of diplomats, foreign employees, clergy, and merchants came from Europe and the U.S. and resided in the settlement. In Japan, the modern postal system didn't work then, and the foreign residents had no choice to communicate with their home country via the consulates. This system is called Services by Consulate. The service scale varied from the handling mail by the consular staff in their spare time during the main business to the systematic one like the British or French PO in Yokohama, which had the independent building and full-time staff for postal service.

I'll describe the history of the location of the French post office in Yokohama, from its opening to closing, and the postmaster Henri Degron, and introduce some mail from/to the settlement.

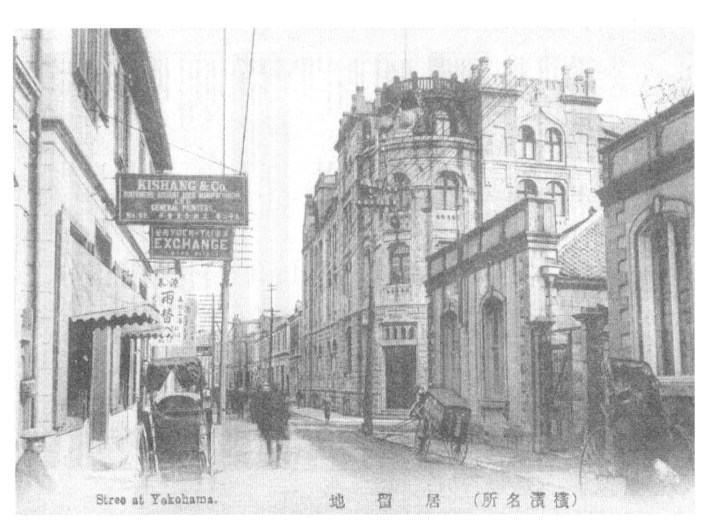

fig.2 居留地本町通 Main Street of the Settlement in Yokohama

1. 在横浜フランス郵便局

1.1 領事館時代を含むフランス郵便局所在地の変遷

1. French Post Office in Yokohama

1.1 History of the Location of the French PO Including the Consulate Period

Period	所在地	Location	図版
1859 – 1862	神奈川宿・慶運寺	Keiunji Temple, Kanagawa-Shuku	fig.3
1862 - 1865	駒形町中横丁	Komagata-Machi Naka-Yokocho	
1865 - 1866	居留地 31 番	Settlement Lot #31	
1866 - 1866	運上所向側	front of the business tax office	
1866 - 1875	本町 5 丁目（別名「弁天」）	Honcho 5 chome (a.k.a. "Benten")	
1875 - 1880	居留地 134 番	Settlement Lot #134	

1.2 開局

　1865 年 7 月 1 日、フランス横浜領事館の郵便事務担当者デグロンが正式に横浜局の局長に任命されました [fig.4]。そして、1865 年 9 月 7 日に横浜入港のフランス郵船デュプレックス（Duplex）号により郵便資材一式が届けられ、この時点で郵政の規定に則った郵便取扱の態勢が整ったことから、同日が開局日と特定されました。

1.2 Opening

On July 1, 1865, Degron, postal service staff in the French consulate in Yokohama, was officially approved as the postmaster of the French PO in Yokohama [fig.4]. The service start day was considered on September 7, 1865. It was because a French mail ship, Duplex, which arrived at Yokohama Port on that day, carried all the materials for handling the mail, and it completed the mail handling system under the postal regulation.

fig.3 フランス領事館 1859 ～ 1862 (慶運寺)
French Consulate (Keiunji Temple)

fig.4 デグロン横浜フランス局長
Henri Degron, the Postmaster of the French PO in Yokohama

5118 Grand Chiffre Losange Cachet à date「5118」大型数字入り菱形抹消印

Cachet à Date YOKOHAMA BAU FRANÇAIS

Cachet àDate YOKOHAMA JAPON

Cachet à Date YOKOHAMA CORR D'ARMÉE

PD（Cahet Payé jusqu'à Destination）

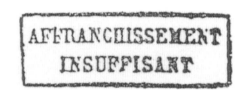

Affranchissement Insuffisant
料金不足印

Cachet 5118 Valeur Déclarée　価格表記印

fig.5 在横浜フランス局使用各種印 Postmarks at the French PO in Yokohama

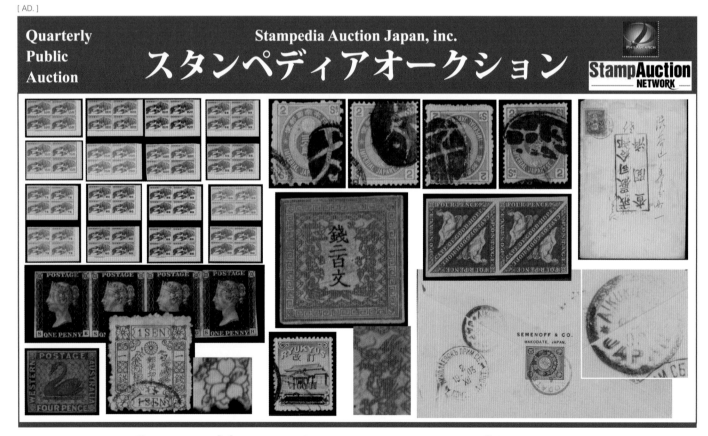

1.4 横浜大火

　1866年11月26日早朝、居留地に隣接する末広町の豚肉商から出火し、西風に煽られて大火災となりました。この大火で日本人市街の3分の2と居留地の5分の1が焼失しました。

　「在庫していた商品が残念ながら先月26日の大火ですべて焼失したことをご報告いたします」[fig.6]。この書簡は、大火後最初に(1866年12月16日)に出航した仏郵船アルフェー号で香港に、さらに別の郵船に積み替えられて1867年2月1日マルセイユ着。その後、陸路で北フランスのラーンスに届けられました。

　当時、フランス領事館(郵便局併設)は運上所向側に移転していましたが全焼しました。「5118」印や郵便切手類が焼失しました。しかし、抹消印以外の日付印や証示印などは焼失を免れたと言われています。新「5118」印2個が翌1867年4月に支給されました。なお、新「5118」印1個の内、「8」が他の印よりも大きく、またその中には十位の「1」が破損している印があります[fig.7]。

1.4 Disastrous Fire in Yokohama

It was in the early morning on November 26, 1866. in Suehiro-Cho, the area next to the settlement, a fire broke out at a butcher shop, and a west wind fanned the flame to make a disastrous fire. This fire burned two-thirds of the Japanese residential area and one-fifth of the settlement.

"We're sorry to inform you that we lost all the merchandise in stock in a fire on 26 of the last month." [fig.6] This letter was the first letter sent after the fire. A French mail ship, Alphée, which left Yokohama on December 16, 1866, carried it to Hong Kong, then another mail ship to Marseille, which arrived on February 1, 1867. From there, it was delivered by land to Reims in North France.

French consulate (with a post office) then, which was moved to the front of the business tax office, lost all the building. They lost "5118" numeral handstamps and postage stamps in a fire. However, the cds or postmarks seemed to escape. Two new "5118" numeral handstamps were provided in April 1867. One of the new "5118" postmarks has the letter "8" larger than others and a broken "1" of the tens digit [fig.7].

fig.6　1866年横浜大火後最初に出港したアルフェー号で逓送。Carried by Alphée, the first ship left Yokohama after the disastrous fire in 1866

fig.7　Broken 1

1-5 本町 5 丁目への移転と大火による臨時措置

　焼失した横浜局の業務は、弁天にあったフランス公使館の隣に移転して再開されました。同地は本町 5 丁目とも表記されています。他方、事務上の扱いとしては、「5118」抹消印が焼失したため、日付印や錨入菱形印が利用された例があります [fig.8]。錨入菱形印は 1867 年 1 月横浜入港のアルフェー号が自船の船舶郵便用を貸与したといいます。なお、この時期、日付印や PD 印が本来の黒や朱に代えて青で押印されている例が散見されます。業務が混乱していたことがうかがえます。

1.5 Move to Honcho 5 chome and Provisional Handling, Caused by the Fire

They continued the works of the post office by moving next to the French legation in Benten, also known as Honcho 5 chome. As for the mail handling, they substituted the cds and the anchor rhombus postmark (Cachet Losange Ancre) for the burnt "5118" numeral handstamps [fig.8]. Alphée, which arrived at Yokohama in January 1867, was said to lend their anchor rhombus postmark normally used for ship mail. At the time, there were many cases of the cds and PD postmarks in blue ink, which should be black or vermilion. The work should have been confused.

fig.8 錨入菱形印
Anchor Rhombus
Postmark

1-6 居留地 134 番

　1875 年 11 月、フランス領事館は山手地区フランス山の麓に移転する計画がまとまり、新領事館落成まで一時的に居留地 74 番に移転しました。これに伴い、領事館と共に仮住まいしていた郵便局も移転することになり、居留地 134 番のデグロン局長の私宅を局舎代わりに利用しました。

1.6 Settlement Lot #134

In November 1875, the French consulate decided to move to Mont France in the Yamate district and provisionally moved to lot #74 until the new building would be completed. The post office also temporarily moved to lot #134, the house of the postmaster Degron, and used there as an office.

1-7. 日本の外国郵便制度の整備

　1871 年 4 月 20 日、東京 - 大阪間に官営郵便路線が開設され、前島密らにより近代郵便制度が発足しました。以来、国内の郵便路線は着実に伸張されていきました。さらに外国との郵便創設を志向する駅逓寮は、英・米・仏の在日局を利用して外国郵便を受発信する仕組みを考案しました。

　横浜以外の日本各地宛到着便は、横浜の各外国局に私書箱を設け、横浜郵便役所が一括して受け取り、駅逓寮によって受取人の居所まで配達する方法を取っていました。この配達料金は国内便扱いとし、先払（受取人払い）であるも

1.7 Development of International Mail System in Japan

The government postal network was established between Tokyo and Osaka on April 20, 1871, and Hisoka Maejima and others built up a modern postal system. Then the postal network in Japan extended steadily. The Postal Service Bureau (Ekiteiryo), which intended to create an international mail system, invented a method of mail to/from foreign countries using the British, U.S., and French post offices in Japan.

Yokohama PO received all the mail to Japan except to Yokohama by establishing PO boxes in each three foreign POs in Yokohama, and then the Postal Service Bureau delivered to the addressee. The delivery rate was for an

のの、2倍料金を徴収しないことも定められました。

他方、外国への差立便は 1872 年 4 月 8 日布告の「海外郵便手続」でその方法を公布しました。すなわち、外国宛の封筒を別の大きめの封筒に入れて、日本国内料金と宛先別に定める外国郵便料金の合計額を日本切手で外側の封筒に貼付し、「外国郵便差出願」と記して駅逓寮に差し出すように定められたのです。

駅逓寮では依頼すべき外国局を選定し、内側の封筒にその局経由、当該在日局に持ち込み外国への送達を依頼するという仕組みでした。因みに、フランス宛書状料金は 15 グラム毎に 32 銭と定められました。当時の在横浜フランス局からフランス本国宛の書状料金は 10 グラム毎に 1 フラン (仏国郵船による) または 1 フラン 30 サンチーム（英国郵船による）でした。

「外国郵便差出願」によって差立てられた実例が現在までに 3 通が確認されています。その内の 1 通は「八戸カバー」の名称で広く知られています。

1-8. 万国郵便連合（UPU）の成立

1870 年代前半まで、世界各国における外国郵便の仕組みは複雑でした。当時の郵便交換条約は当事者 2 国間で締結されることが多く、料金その他の扱いにおいて統一性を欠いていて、利用者には不便なものでした。そこで万国共通の郵便条約の構想が浮上し、1862 年 8 月にアメリカの郵政長官ブレアが郵便国際会議を提唱しました。

翌 1863 年、主要 15 カ国の代表がパリに参集して郵便物の重量と料金の統一、乗継料金の簡易化などが議論されました。この会議は 1863 年パリ小会議と称され、後年の万国郵便連合の萌芽とされています。

1874 年、ベルリンに 22 カ国の代表が集まり、最初の郵便大会議が開催されました。そして同年 10 月 9 日、「一般郵便連合（GPU）の創立に関する条約」が調印され、次回のパリ大会議において万国郵便連合への改称が議決されました。当初の加盟国は 22 カ国で、条約の発効日は 1875 年 7 月 1 日と定められました。フランスは、国内事情から 1876 年 1 月 1 日からの加盟となりました。日本の正式な万国郵便連合への加盟日は 1877 年 6 月 1 日で、28 番目の加盟国でした。万国郵便連合への加盟によって、6 月 20 日以降、加盟国宛の書状料金は 10 銭と定められました。ただし、アメリカと上海宛は特別料金 5 銭のままでした。以後、再三料金は改定されました。

1870 年 12 月 21 日公告、1871 年 1 月 1 日適用開始ではブリンディシ経由の料金が設定されましたが、これは普仏戦争のためで、英国船はマルセイユに代えてイタリア南端のブリンディシに母港を置くことになり、イタリアは通過料を徴収しました。従って、英国船による逓送の場合、通過料を上乗せした料金を設定しました。

また、1877 年 3 月 16 日公告、4 月 1 日施行、5 月 30 日適用開始では GPU 加盟国間の郵便料金は大幅に引下げられました。

internal mail. The internal rate was charged to the recipient, but it was not doubled.

The "International Mail Handling Procedure," decreed on April 8, 1872, announced the method to mail to foreign countries. Enclose a letter to a foreign country in another bigger sized envelope, stick the Japanese stamps of total postage of internal and international rate on the outer envelope, handwrite "international mail subscription" and send it at the Postal Service Bureau.

The Postal Service Bureau chose the forwarding foreign post office, put the stamps of the country of destination, and mailed it at the assigned foreign post office in Japan. The letter rate to France was 32 sen per 15g, and from the French PO in Yokohama to the home country France was 1 Fr. per 10g by a French mail ship or 1.30 Fr. per 10g by a British mail ship.

There are three recorded items sent by "international mail subscription." One of them is well known as "Hachinohe Cover."

1.8 Establishment of Universal Postal Union (Union Postale Universelle [UPU])

The international mail system of each country was complicated until the first half of the 1870s. Generally, the mail exchange treaty was concluded between the two concerned countries, and it wasn't convenient for the users because of the lack of standardization of postage and other manners of handling. Then, the idea of uniformed postal treaty appeared, and in August 1862, U.S. Postmaster General Blair called for an international postal congress.

In 1863, the next year, the delegates of the leading 15 countries met in Paris and argued about standard weight and postage rate of the mail, simplification of transit, etc. It is called as Paris Congress of 1863. That was the first attempt of the establishment of the UPU in the future.

In 1874, the delegates of 22 countries met in Berlin. It was the first postal grand congress. On October 9 of the same year, the "Treaty concerning the formation of a General Postal Union (Union Générale des Postes [GPU])" was concluded. At the next grand congress in Paris, it was renamed the Universal Postal Union. The members were 22 countries then, and the effective date was scheduled for July 1, 1875. France participated in the GPU on January 1, 1876, for internal affairs. Japan officially joined on June 1, 1877, as the 28th member. Since Japan became a GPU member, the letter rate to the member countries was 10 sen after June 20. However, the special rate of 5sen, the rate to the U.S. and Shanghai, was still working. After then, the postage rate was revised repeatedly.

The notification on December 21, 1870, which was applied on January 1, 1871, set the rate via Brindisi caused by the Franco-Prussian War. British ships set a mother port in Brindisi, on the southern end of Italy, instead of Marseille, and Italy charged the transit fee. So, the rate for the delivery by a British ship was including an additional transit fee.

And the notification on March 16, 1877, which was enforced on April 1 and applied on May 30, the rate of the GPU member countries decreased drastically.

1-9. 閉局

　1877 年になるとフランス横浜局発着の郵便物は急激に減少しました。後期（1877～1880 年）の現存郵便物は 200 通未満と推定されており、開局から閉局まで（1865 ～ 1880 年）の同局発着郵便物の推定現存総数（約 3,000 通）の 7% でしかありません。

　フランス本国の郵便電信省郵政総局は郵政月報第 23 号（1880 年 3 月）で「フランス横浜郵便局は 1880 年 4 月 1 日以降、その業務を停止する」と告示しました。デグロン局長も同主旨の公告を 3 月 6 日付 L'Écho du Japon 紙に掲載しました。「4 月 1 日以降、従来フランス局で行ってきた郵便物の取り扱い業務はすべて日本郵政当局により実施される」。

　1865 年 9 月 7 日の開局以来、14 年 7 カ月に及ぶ同局の歴史に終止符が打たれました。なお、デグロンは閉局後もフランス郵政上の地位としては横浜局長のまま、1 年間は日本の駅逓局顧問として二重に雇用されていました。この間、デグロンはフランス郵便局の所在地であった居留地 134 番に居住し続けました。フランスのルボン大佐夫人宛の書状が不在のため、フランス横浜局（居留地 134 番―デグロン私宅兼フランス局）宛に戻されたカバー [fig.9] を紹介します。

1.9 Close the PO

The mail to/from the French PO in Yokohama decreased in 1877. The number of items between 1877 and 1880 is estimated less than 200, which is just 7% of the total possible amount (approx. 3,000 items) at the PO from open to close (1865 - 1880).

Ministry of Posts and Telecommunications (Ministère des Postes et Télécommunications) in the home country France notified in the Monthly Bulletin No.23 (Bulletin Mensuel No.23), published March 1880, "the French PO in Yokohama will stop the operation from April 1, 1880." Postmaster Degron announced the same notification on the newspaper L'Écho du Japon dated March 6. "From April 1, all the mail handling at the French PO will be operated by the Japanese Postal Authority."

It was the end of the 14-years-and-7-months history of the French PO from the opening of on September 7, 1865. However, Degron was still a postmaster of the French PO in Yokohama after the close under the French Postal Authority, and double employed for a year as an adviser at the Japanese Postal Service Bureau. Degron continued to live on lot #134, where the French PO was. Here I show a returned letter cover to the French PO in Yokohama (lot #134, Degron's residence, and French PO) caused by the absence of the addressee, Mrs. Colonel Lebon in France [fig.9].

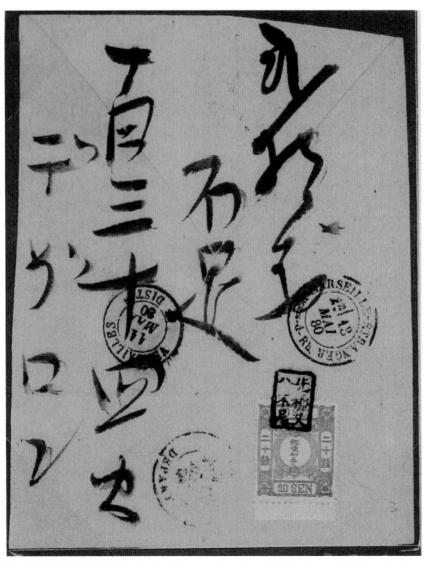

fig.9 名宛人不在でフランス横浜局に戻されたカバー Returned to French Yokohama PO, by the absence of the addressee

2. 「デグロン君カバー」

2-1 デグロンの生立ち

　前掲の在横浜フランス局のデグロンは 1839 年 4 月 14 日、ヴェルサイユ市内で誕生しました。父親は司法書士、母親は近郊クレピエールの富農の娘で、一人っ子のアンリは不自由のない暮らしを送っていたということです。しかし、彼が 14 歳の時、父親は投獄されてしまいました。理由は不明。この事件のためか、デグロンは高等教育を受けていません。数回にわたり外務省の書記官試験を受けましたたが合格しなかったということです。

　1861 年 2 月 2 日もしくは 3 日、フランス軍艦 ラ・ドルドーニュ号の乗組員として来日しました。その後、1862 年 1 月末から 8 月初めの間に同艦を降りて公使館付の書記官補に任命されました。

　在上海局は 1862 年 12 月 19 日に開設されました。同局には大型数字 5104 入り菱形印や日付印が配付されました。デグロンは郵便実務習得などの目的で 1864 年 1 月に上海局に出張しました。上海から帰任した直後の 2 月 4 日付の Daily Japan Herald 紙に初めてデグロン名義で郵便受付の公告が掲載されました。デグロンの肩書は郵便取扱人となっています。

　デグロンはすでに述べた通り、1865 年 7 月 1 日、本省から横浜局の局長に任命されました。そして 1880 年 3 月 31 日の廃局まで局長の任にありました。デグロンは廃局後 1 年間、駅逓局長、前島密の顧問として駅逓局雇となりました。そして、1881 年 4 月 24 日フランス郵船タナイス号で 20 年余り滞在した日本に別れを告げ、クレピエールに帰郷しました [fig.10]。その後も植物の調査などで再三訪日し、日仏文化交流に大きな功績を残しました。

2. "Degron-Kun Cover"

2-1 Degron's Early Years

Henri Degron, Postmaster of the French PO in Yokohama, was born in Versailles on April 14, 1839. His father was a judicial scrivener, and his mother, a daughter of a wealthy farmer in Crespières nearby. Their only son Henri lived in comfort. However, in his 14 years old, his father was imprisoned for an unknown reason. Probably that event impeded him from receiving higher education. He tried sometimes taking the examinations for secretary of the Ministry of Foreign Affairs, but never passed.

On February 2 or 3, 1861, he came to Japan as a staff of a French warship, La Dordogne. Then, between the end of January and early in August 1862, he landed and was assigned the secretary's assistant of the legation.

The French PO in Shanghai opened on December 19, 1862. At the PO, the rhombus postmark with a large-size number "5104" (Losange Grand chiffre 5104) and the cds (Cachet à Date) were supplied. Degron had a business trip to Shanghai PO to learn the mail handling work in January 1864. And on February 4, just after the trip, the announcement of the mail acceptance under his name appeared for the first time in the newspaper Daily Japan Herald. His status then was a mail handling officer.

As I described above, The postal authority appointed Degron postmaster of the French PO in Yokohama on July 1, 1865. He was in that position until on March 31, 1880, when the PO closed. For one year after the close, Degron was employed by the Postal Service Bureau (Ekiteikyoku) as an advisor to the director, Hisoka Maejima. He left Japan, where he lived more than twenty years, by a French mail ship, Tanaïs, on April 24, 1881, and went back to Crespières [fig.10]. He visited Japan repeatedly after then to study the plants, etc. He had remarkable achievements in cultural exchange between France and Japan.

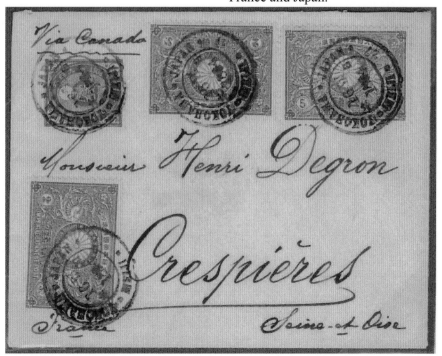

fig.10 故郷クレピエールに帰国後のデグロン宛カバー to Degron, after Back to Crespière

2-2「デグロン君カバー」

「デグロン君カバー」とは、幕末から明治初年にかけて、在横浜フランス局を経由して、東京などからフランス宛に送られた郵便物で、日本とフランス両国の郵便切手が混貼された郵便物を指します。

横浜以外の日本各地（殆どは東京ですが、最近、フランスの技術で建設され、技術指導を受けた在富岡の製糸工場発信のカバーが発見されました）から、日本の国内便によって、先ずはフランス横浜局・局長デグロン宛に配達され、フランス横浜局から先はフランスの郵便物として本国の宛先に送達されました。

日本とフランスの異なる郵便区間を通過するため、それぞれの区間に有効な切手が混貼された郵便物です。日本国内の宛先、すなわちフランス横浜局・局長デグロンとフランス国内の最終宛先と二通りの宛先が記された郵便物と見ることもできます。

2-3「デグロン君カバー」の出現

1858 年 4 月、「シャナハン・オークション」カタログに日本の手彫切手とフランス切手が同じ封筒に貼られているカバーが掲載されました。後に言う「デグロン君カバー」の郵趣界への初めての出現でした。参考値は 80 英ポンドでした。しかしながら、5 月 3 日のシャナハン本社（アイルランド首都ダブリン）での競売では無入札に終わりました。

同カバーは 5 か月後の 10 月 4 日の競売に再出品されました。今回は別のデグロン君カバー 1 通も出品されました。60 ポンドと前回より 25% 評価を下げています。新たに出現したカバーは 50 ポンドの評価でした。当時の為替レートや日本の給与水準を考慮すれば現在の 80 万円（7 300 米ドル）に相当します。落札値はさらに高値になる可能性もあり、また競売人の手数料を加算しなければなりません。

1964 年、「国際郵便切手展」が全日本郵趣連盟主催、朝日新聞社後援で東京と大阪の両会場で開かれました。同展には「デグロン君カバー」2 通が展示されました。実物の一般初公開でした。

2-4「デグロン君カバー」の性格

「デグロン君カバー」方式は、長いこと陸軍省雇のフランス軍事団員に対する優遇措置と考えられていました。

しかし、現在では「デグロン君カバー」方式は優遇措置など特別の郵便取扱方法ではなく、普通の郵便規則で説明可能な通常の郵便物と考える方が妥当だと言われています。すなわち、郵便物上の仏文の宛名とフランス切手を無視すれば、宛名は和文で明記され、料金が日本切手で完納された国内郵便物です。また、仏文の記載やフランス切手の貼付を禁止する規定は「郵便規則」にも「差出人心得」にもありません。

2-2 "Degron-Kun Cover"

"Degron-Kun cover" is the mail from Tokyo etc. via the French PO in Yokohama to France, from the end of the Edo Era to the early Meiji, which was franked by both Japanese and French stamps.

The first step of this international mail method was mailing via the internal system to Postmaster Degron of the French PO in Yokohama. It was from everywhere in Japan but Yokohama (mainly from Tokyo, but recently we've found a cover sent from Tomioka Silk Mill, which French technology helped to construct and where French silk production skill was introduced). Then it was sent as a French mail to the addressee in France.

Since the mail transited the different postal zones, which were of Japan and France, the effective stamps of each zone were put on it and shows two addressees; the addressee in Japan, Degron, Postmaster of the French PO in Yokohama, and the addressee who received it in France.

2-3 Appearance of "Degron-Kun Cover"

A cover bearing Japanese engraved stamps and French stamps appeared in the Shanahan Auction Catalogue in April 1858. It was a philatelic debut of "Degron-Kun cover." The reference value was £80 (Sterling pound). However, nobody made a bid for it at the auction in the main office of Shanahan on May 3 (in Dublin, Capital of Ireland).

That cover was displayed again at the auction on October 4. At the auction, another Degron-Kun cover also appeared. The value was £60, decreased by 25% comparing with the previous sale. And another cover valued at £50. Considering the exchange rate and the standard salary in Japan then, it equivalents to 800,000 YEN now (US$7,300). The hammer price would be possibly higher than the value, and also, it needed to pay the premium for auctioneer.

In 1964, the "World Stamp Exhibition" was held in Tokyo and Osaka, promoted by All Japan Philatelic Federation and sponsored by the Asahi Shinbun Company. Two "Degron-Kun covers" were shown in the Exhibition. It was the actual articles' debut in public.

2-4 Feature of "Degron-Kun Cover"

We had thought for a long time that the "Degron-Kun cover" method was a courtesy for French officers of the Military Mission of Armies (Mission Militaire des Armées).

However, now we think it is appropriate that it was not a particular handling method like a courtesy for military, but the ordinal mail handling which a common postal regulation could explain. It means, if you ignore the addressee in French and French stamps, the mail was just an internal mail in Japan, which had an addressee in Japanese, franked by Japanese stamps. And nor "Postal Regulations" nor "Rules of Sender" included the prohibition for description in French or using French stamps.

他方、フランス横浜局でも、宛名は仏文で明記されているし、料金は自国切手で前納されているので引受けを拒否する理由はありませんでした。日本切手と和文デグロンの宛名も横浜以遠の逓送上問題はありませんでした。

この「デグロン君カバー」方式は1873年1月13日差出から1877年12月3日差出まで現存例があり、5年間の長期にわたって継続実施されていて、公認の郵便物取扱方法であったと推察されます。従って、この「デグロン君カバー」の開始に当たっては、日本の官庁間の調整も、日仏両郵便局間の協議も新しい規則を制定する必要もなかったと結論付けられます。

2-5「デグロン君」印のタイプ分類

東京からフランス横浜局まで国内郵便として送るための宛名、デグロン局長の役職と所在地を記した木製の印判は日本人の印判屋に作らせてフランス軍人宿舎である教師館に配備されていました。現在、下記6タイプが知られています [fig.11]。

On the other hand, at the French PO in Yokohama, since the mail had the addressee in French, franked by French stamps, they had no reason to deny accepting it. Japanese stamps and the addressee Degron in Japanese didn't have a problem with delivery beyond Yokohama.

There are remaining items of the "Degron-Kun cover" method from January 13, 1873, to December 3, 1877. Since it worked for such a long time as five years, it was probably accepted as a proper mail handling method. So, I suppose when they started the "Degron-Kun cover" system, they didn't need any adjustment among the authorities concerned in Japan, a conference between French and Japanese post offices, or establishment of a new rule.

2-5 The Classification of the "Degron-Kun" Seals

Wooden address seals of Postmaster Degron, which were used to deliver from Tokyo to the French PO in Yokohama, were made by Japanese seal shops and deployed to the foreign teachers' residences where French soldiers stayed. We know the six types below by now [fig.11].

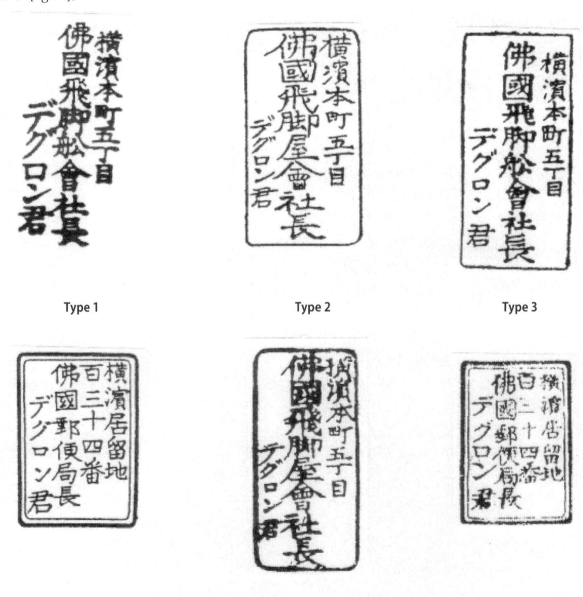

Type 1　　　　　　Type 2　　　　　　Type 3

Type 4　　　　　　Type 5　　　　　　Type 6

fig.11 デグロン君印　タイプ1～6 Degron-Kun Seals, Type 1 to 6

2-6「デグロン君」印押捺カバー

　一般に「デグロン君カバー」はフランス横浜局の所在地とデグロン君が和文で刻印された刻印された印判が押捺され、日仏両国の郵便切手が混貼されているカバーと定義されていることは前述の通りです。

　現実には日本切手が剥がされ、フランス切手だけがカバー上に残っている、いわば損傷「デグロン君カバー」も少なからず現存します。これはフランス到着後、日本切手が珍しく名宛人もしくは第三者が剥がしたのかも知れません。また、fig.11-1 に示すように日本切手をカバーから剥がした痕跡はありませんがフランス切手のみが貼付されているカバーも少数あります。これらは差出人もしくは代理人が直接フランス局に持参して出状したものと思われます。

　日仏郵便史研究の第一人者・松本純一氏の調査における最新のデータベースによれば、宛名手書きのカバーも含め 105 通が確認されています。内訳は下記の通りです。

2-6 Cover with "Degron-Kun" Seals

As I described above, generally, a "Degron-Kun cover" had the seal of the address of the French PO in Yokohama and Degron-Kun's name in Japanese and the stamps of both France and Japan.

There are many damaged "Degron-Kun covers," which Japanese stamps were peeled and only had French stamps. It is probably because the addressee or someone else removed the rare Japanese stamps in France. Also, a small number of covers only bearing French stamps without any trace of removing Japanese stamps (see fig.11-1). Those were supposed to be sent by the sender or his/her deputy directly at the French PO.

According to the latest data investigated by Mr. Junichi Matsumoto, the leading expert of French-Japanese postal history, 105 items, including handwritten addressee, have been recorded. The details are as follows:

印判タイプ Seal Type	完全カバー Entire cover	フロント Front	日本切手脱落他 with missing stamps	日本郵便不関与（日本切手不貼付） without via Japanese post (without Japanese stamps)
1	53		33	1
2a	1			2
2b	1			
3	1			
4			1	
1+4			1	
5	6			2
6	1	1		
宛名筆書 Hand written addressee	2			

fig.11-1

JAPAN STAMP AUCTIONS

★ オークションスケジュール ★

フロア	メール	フロアセール(土・日)	メールセール(火)	東京下見会(日)	ご出品締切(土)
第115回	第100回	12月5日〜6日	12月8日	11月22日	―
第116回	第101回	2021年 2月27日〜28日	3月2日	2月21日	12月19日
第117回	第102回	5月29日〜31日	6月1日	未定	3月19日(金)
第118回	第103回	国際展開催予定のため　詳細未定			

オークションについて

　オークションは大阪で年4回開催します。各回共、幅広い日本及び関連地域の中高級品のみを扱います。4,000〜6,000ロットの出品です。オークション見本誌をご希望の方は、フロア・メール2冊で300円(切手代用可)でお届けします。

■ オークション会場は、大阪駅前第3ビル17階です。

■ 東京下見会は、第115回・第116回は切手の博物館3階で開催します。

■ オークション出品物の記事は、オークション誌発行の10日前にホームページにてご覧いただけます。また、ダウンロードでの先行下見も可能です。

年会費

1年間(1月〜12月迄) 2,000円

◇◇◇ 途中入会による年会費 ◇◇◇

入会月による年会費

1・2月にお申込▶	2,000円(年末迄)
3・4月にお申込▶	1,600円(年末迄)
5・6月にお申込▶	1,200円(年末迄)
7・8月にお申込▶	800円(年末迄)
9〜12月にお申込▶	2,400円(翌年の年末迄)

ご出品について

1点 3,000円程度以上の日本関連品限

● ご出品は上記期日必着でお送り下さい。
（定形以外の郵便物は私書箱ではなく事務所へお送りください。）

● 弊社の判断で掲載基準に達しないものはお返ししております。記事の編集は当方でいたしますので、詳細リストは不要です。

● 多数のご出品物をお預かりしており、編集能力の限界を超えております。大量出品等の場合、数回にわたっての掲載になる場合もございます。

● 大阪開催のフロアセールと併催しております【大阪駅前第3ビルバザール】の開催は現時点では確約できません。COVID-19の感染状況により流動的です。確定し次第、ホームページ等でご案内いたします。

ジャパン・スタンプ商会

通信先 〒530-8691 大阪北郵便局私書箱89号　Tel.06-6347-1601/Fax.06-6347-1602
www.japan-stamp.com　E-mail : japan-stamp@juno.ocn.ne.jp

事務所 〒530-0001 大阪市北区梅田1-1-3　大阪駅前第3ビル14F1号
ゆうちょ振替：00980-3-51454　取引銀行：池田泉州銀行 石橋支店 当座 #85749
営業：火〜土 10時〜6時(日・祭・月曜定休)

3. 横浜外国人居留地専有者の変遷とその郵便物

3. History of the Owners of Foreign Settlement in Yokohama and their Mails

3-1 改正新条約

居留地は不平等の規定が多々見られる通商条約を全面的に見直して改定新条約が発効する 1899 年まで存続しました。新条約締結後も同地で営業を続ける外国商館や外国人も多く存在しました。居留地内の土地や建物の専有者は、理由はさまざまですが、ひっきりなしに入れ替わるのが常でした。

3-1 Treaty Revision

The foreign settlements remained until 1899, when the new treaty, which was established by revising the unequal commercial treaty, became effective. Even after the conclusion of the new treaty, there were many foreign residences and merchants continuing business. The owners of land or building in the settlement frequently changed for various reasons.

3-2 居留地 80 番の専有者の変遷

本稿では、ペィル兄弟が 1875 年 12 月 17 日に創業し、店舗と住居を併設していた居留地 80 番、84 番と 85 番の 1868 年から 1899 年までの専有者の変遷を一覧表に纏め、文末に掲載し、さらに専有者の受発信郵便物を紹介します。

ここで、鳥居 民・著「横浜山手・日本にあった外国」の一部を引用させていただきます。

> 居留地を訪れる物見高い見物客は商館の庭先に入り込み、棒を持った南京人が現れれば、いたずらっ子のように声をあげて逃げだした。ところが 80 番の建物は、見物客が入り込んでも怒る人がいなかった。そっと石段を上がり、円柱のある玄関ポーチに足を踏み入れても、どこからも怒鳴り声は聞こえてこなかった

これはカソリック教会（専有期間 1862 ～ 1906）を指しています。

3-2 Owners of Lot #80

I show at the end of the article a table including the owners from 1868 to 1899 of lot #80, #84, and #85. Those three places were where the Peyre brothers started business on December 17, 1875, and had their shop with residence there.

Now I describe the owners' mail below, and here I refer to some sentences of "Yamate, Yokohama, Foreign Country Inside Japan," written by Tami Torii.

> *Curious people visiting the settlement got into the garden of the trading house. And when a Western came to drive them with a pole, they escaped shouting like naughty children. However, in the building on lot #80, there was nobody to get angry with the sightseers. Although they went up the stone steps slowly and went into the columned entrance porch, they didn't hear any shout.*

It was Roman Catholic Church (1862 - 1906).

fig.12 ペィル兄弟洋菓子店 (居留地 80 番) から父ジャン宛 長兄アルチュール 1876 年 4 月 10 日差立
Sent by the Oldest Son Arthur on April 10, 1876, from Peyre Brothers' Patissiers and Confectioners (Lot #80) to Their Father Jean

80番は本町通りと本村通りに、84番と85番は本町通りと長崎町に挟まれた細長い土地で、同じ地番に複数の建物が軒を連ねていました。

1875年12月から1878年6月まではペィル兄弟洋菓子店、1887年から1889年の間はカソリック教会、ラヒンカーン商会、スイス時計商会などが80番に所在していました。

Those were strips of land. Lot #80 was between the roads of Honcho and Honmura, and lot #84 and #85 were between Honcho and Nagasaki-Cho. Some buildings were constructed in the same lot.

The buildings on lot #80 were Peyre Brothers' Patissiers and Confectioners (Peyre Frères Pâtissiers et Confiseurs, December 1875 to June 1878), Catholic Church (1887 – 1889), Rahimkhan & Co., Depot of Swiss Watches, etc.

a. ペィル兄弟洋菓子店

(専有期間：1875〜1878年：80番)

南フランス、プロヴァンス地方の寒村ムーリエスの豪農だったペィル一族の中にペィル＝ジャンがいました。彼は4男1女の父親でした。長男ジャン＝アルチュール、次男マチウ＝ウジェンヌ、三男サミュエル＝ポール、四男ジュール＝ダニエル、長女エリーゼです。

1874年4月21日以前に先ず三男が横浜にやって来て、1872年にボナが84番で開業したオリエンタル・ホテルの料理長補佐として採用されました。その後、1875年12月12日、長男がフランス郵船のヴォルガ号号で来浜するのを待って、12月17日に80番で「ペィル兄弟洋菓子店」を創業しました。長男アルチュールが80番の洋菓子店から実家の父親に宛てたカバーです [fig.12]。

1878年、84番所在オリエンタル・ホテルのオーナーだったボナは海岸通りの「グランド・ホテル」を手に入れるとオリエンタル・ホテルをペィル兄弟に賃貸しました。彼らは「ペィル兄弟ホテル」と改称し、洋菓子店と喫茶店も併設しました。84番の同ホテルから実家宛カバーです [fig.13]。

a. Peyre Brothers' Patissiers and Confectioners

(1875 - 1878 at the lot.#80)

Peyre Jean was a family member of the wealthy farmer in Mouriès, in South France. He had four sons and a daughter. They were the oldest son Jean Arthur, the 2nd Mathieu Eugéne, the 3rd Samuel Paul, the 4th Jules Daniel, and the daughter Elise.

Before April 21, 1874, the 3rd son came to Yokohama first and got a job as an assistant of the chief cook in Oriental Hotel, which L.Bonnat opened on lot #84 in 1872. On December 12, 1875, the oldest son came by a French mail ship (Messageries Maritimes), Volga, and they started "Peyre Brothers' Patissiers and Confectioners" on lot #80 on December 17. fig.12 is a cover, which the oldest Arthur sent from the shop on lot #80 to his father.

In 1878, Bonnat, the owner of the Oriental Hotel on lot #84, obtained the "Grand Hotel" in Kaigan Dori and lent Oriental Hotel to the Peyre brothers. They renamed it "Peyre Brothers' Hotel (Hotel Peyre Frères)," with confectionery and cafeteria. fig.13 is a cover sent from the hotel on lot #84 to their family.

fig.13 ペィル兄弟ホテル (居留地 84 番) からペィル父子宛 三男サミュエル 1878 年 8 月 7 日差立
Sent by the 3rd Son Samuel on August 7, 1878, from Peyre Brothers' Hotel (Lot #84) to Their Family in France

1882 年 8 月 8 日未明、83 番から出火、ペイル兄弟ホテルも類焼したため、ホテル業から撤退を余儀なくされ、同年 12 月 4 日に隣接する 85 番でペイル兄弟洋菓子店を新装開店しました。

なお、洋菓子店経営の傍ら、次男ウジェンヌは日本郵便切手総代理店の名称で当時の日本切手をフランスを中心にヨーロッパ諸国に輸出していました [fig.14]。

ペイル兄弟は改正新条約の発効を機に故郷ムーリエスに帰郷することを決意し、1899 年 3 月 4 日、北フランスのアルザス地方出身のドイツ系フランス人ジュスタン＝ヴェイユに洋菓子店と併設の喫茶店を譲渡しました。ヴェイユは 1911 年に廃業するまで、ペイル兄弟洋菓子店の看板を下ろすことはありませんでした。

Before the dawn on August 8, 1882, a fire broke out at lot #83, and Peyre Brothers' Hotel caught fire. They had to give up the hotel business, and on December 4 of the same year, on lot #85, next to the hotel, newly opened Peyre Brothers' Patissiers and Confectioners.

The 2nd son, Eugéne, did another business besides the management of the confectioner. He exported Japanese stamps to Europe, mainly France as a general commission agent of Japanese stamps [fig.14].

When the new treaty became active, the Peyre brothers decided to come back to their hometown Mouriès. On March 4, 1899, they transferred their confectioner and cafeteria to Justin Weil, who was a German-French from the Alsace region in northern France. Weil never changed the goodwill of Peyre Brothers' Patissiers and Confectioners until he closed the shop in 1911.

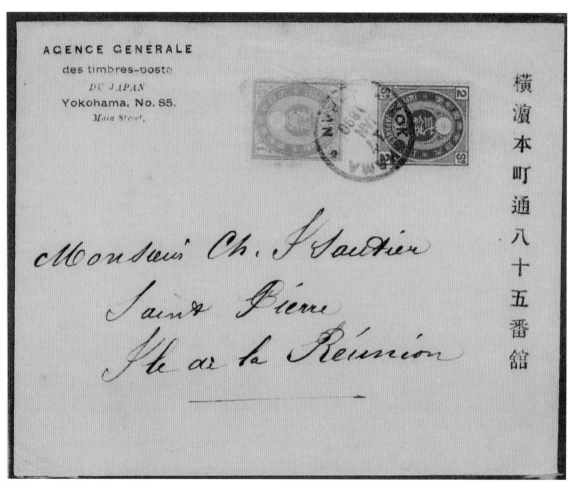

fig.14 日本郵便切手総代理店 (居留地 85 番) からレユニオン島宛 次兄ウジェンヌ 1890 年 1 月 17 日差立 Sent by the 2nd Son Eugéne on January 17, 1890, from the General Agent of Japanese Stamps (Lot #85) to Réunion Island

b. ラヒンカーン商会

（専有期間：1887 〜 1889 年）

インド人ラヒンカーン商会宛。伊勢屋吉次郎 1888 年 12 月 10 日発信はがき（和文）[fig.15]。なお、同商会の詳細は不明。

b. Rahimkhan & Co.

(1887 - 1889)

Postcard from Yoshijiro ISEYA to an Indian Rahimkhan & Co., on December 10, 1888 (in Japanese) [fig.15]. The detail of the company was unknown.

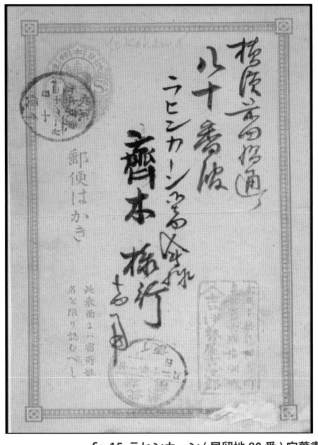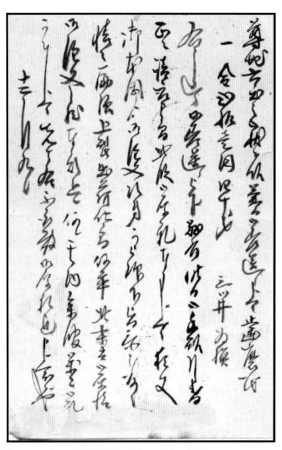

fig.15 ラヒンカーン (居留地 80 番) 宛葉書 Postcard to MM. Rahimkhan (Lot #80)

c. スイス時計商会

（専有期間：1892 〜 1900 年）

スイス時計商会はシュナイダーが経営する時計、宝石、眼鏡製造会社で、1892 年に 80 番で創業しました。在神戸のバチスト = ランジュ発信のシュナイダー宛葉書（仏文）[fig.16]。

c. Depot of Swiss Watches

(1892 - 1900)

Depot of Swiss Watches was a manufacture of watches, jewelry, and glasses, managed by Geo. Schneider, which opened on lot #80 in 1892. fig.16 is a postcard sent by Baptist Runge in Kobe to Schneider (in French).

fig.16 スイス時計商会 (居留地 80 番) 経営者 シュナイダー宛 To Geo. Schneider of Depot of Swiss Watches (Lot #80)

d. スゾール商会

（専有期間：1809 〜不明)

　スゾール商会はエンジンやボイラー等の機械類、ミシュランのタイヤ、ワイン他酒類の輸入販売店であり、ユニオン火災保険会社代理店でもありました。居留地廃止後に 80 番で開業しました。同社発信リヨン宛カバーです [fig.17]。

d. L. Suzor & Co.

(1809 - unknown)

L. Suzor & Co. was an import and sales shop of the machinery such as engine or boiler, Michelin's tire, and wine and other alcohol, and also the commission agent of Union Fire Insurance Co., Ltd. After the end of the settlement, he started business on lot #80. fig.17 is a cover from there to Lyon.

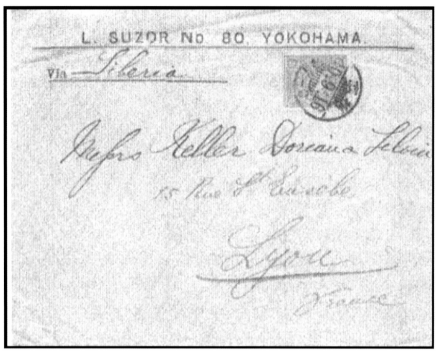

fig.17 スゾール商会 (居留地 80 番) 差立リヨン宛 From L.Suzor (Lot #80) to Lyon

fig.18 ジャクモ (居留地 80 番) 差立リヨン宛 From J.M. Jacquemot (Lot #80) to Lyon

e. ジャクモ

（専有期間：1869 〜 1871 年）

スイス人ジャクモは開港直後の 1860 年あるいは 1861 年、まだスイスと日本とで通商条約が締結されていない時代に英国人として来日しました。

当時、ジャクモは英国人の中で最高齢の 45 歳で、絹検査官、貿易商、保険代理業に携わっていました。当初、82 番に事務所と住居を構えていましたが、1869 年には住居だけ 84 番に移しました。

1872 年ボナが 84 番にオリエンタル・ホテルを開業すると、住居を 82 番に戻しました。その後 1877 年に離日するまで 82 番で営業を続けました。同事務所から発信した書簡（仏文）です [fig.18]。

e. J.M. Jacquemot

(1869 - 1871)

Jacquemot from Switzerland came as a British in 1860 or 1861 when Switzerland and Japan didn't conclude a commercial treaty.

Jacquemot was the oldest British then, 45 years old, and worked as a silk inspector, a trader, and an insurance agent. At first, he had an office and residence on lot #82, and the only residence was moved to lot #84 in 1869.

In 1872, when Bonnat opened Oriental Hotel on lot #84, he returned the residence to lot #82. He continued business on lot #82 until 1877 when he left Japan. fig.18 is a letter from the office (in French).

f. オリエンタル・ホテル

（専有期間：1872 〜 1878 年）

ペイル兄弟三男サミュエル＝ポールは 1874 年 4 月 21 日以前に（正確な期日不明）横浜に到着し、1872 年にボナが 84 番で開業したオリエンタル・ホテルに料理長補佐として採用されたことは既に述べました。サミュエルがムーリエスの父親に宛てたカバーです [fig.19]。

f. Oriental Hotel

(1872 - 1878)

The 3rd son of the Peyres, Samuel Paul, came to Yokohama before April 21, 1874 (accurate date unknown), and as I described before, he was employed as an assistant of the chief cook of Oriental Hotel, which Bonnat opened on lot #84 in 1872. fig.19 is a cover sent by Samuel to his father in Mouriès.

fig.19 オリエンタル・ホテル (84 番）　から父親ジャン宛 三男サミュエル 1875 年 11 月 13 日差立
Sent by the 3rd Son Samuel on November 13, 1875, from Oriental Hotel (Lot #84) to His Father Jean

g. ペイル兄弟ホテル

（専有期間：1878 〜 1882 年）

ペイル兄弟は 1878 年にボナからオリエンタル・ホテルを賃借し、「ペイル兄弟ホテル」と改称し同年 6 月に開業しました [fig.20]。

g. Peyre Brothers' Hotel / Hôtel Peyre Frères

(1878 - 1882)

The Peyre brothers rented Oriental Hotel from Bonnat in 1878 and opened it in June of the same year, renaming "Peyre Brothers' Hotel" [fig.20].

fig.20 ペイル兄弟ホテル (84 番) 道路左側 Hotel Peyre Frère (Lot #84) on the Left of the Street

h. 横浜英和女学校

（専有期間：1885 〜 1888 年）

1880 年、横浜山手外国人居留地 48 番でブリッタンがプロテスタント系ミッションスクール「ブリテン女学校（英和女学校）を開校しました。1883 年に山手 120 番に移転しました。

1886 年には「メソジスト・プロテスタント・ミッションスクール」（横浜英和女学校）と改称し居留地 84 番に移りましたが、1889 年にはまた山手に戻りました。1885 年からミス・マーガレット・ブラウンが校長を務めました。その後、生徒数が増えたため 1916 年に丘一つ向こうの蒔田に移りました。

なお、1939 年戦時色が強くなり、「英和」という敵国の名前が校名とは不謹慎という風潮の中で「成美学園」と名を改めました。1996 年「横浜英和女学院」と旧名に復しましたが、その後「英和女子学院」とさらに改名され現在に至っています。本町 84 番に移転後の校長、ミス・ブラウン宛の葉書です [fig.21]。

h. Methodist Protestant Mission School in Yokohama

(1885 - 1888)

In 1880, Brittan established a protestant mission school, "Britain Girls School" (Eiwa Girls School), on lot #48 of the foreign settlement in Yamate, Yokohama. The school was moved to lot #120 in Yamate in 1883.

The school was renamed in 1886 "Methodist Protestant Mission School" and moved to lot #84. But in 1889, it came back to Yamate. Miss Margaret Brown worked as a director from 1885. Then, by increasing the number of students, it was moved to Maita, beyond the hill, in 1916.

The school name in Japanese means "English-Japanese". So, in 1939, under a wartime feel, people thought the school name was inconsiderate, then, the school was renamed "Seibi Gakuen." The name was changed back to "Yokohama Eiwa Girl's School" in 1996, then renamed again "Eiwa Girls School," its present name. fig.21 is a postcard from the director Miss Brown after the move to Honcho lot #84.

i. フランス領事館

（専有期間：1889 ～ 1894 年）

フランス領事館は火災などの理由で下記のように移転を繰り返しました。

i. French Consulate / Consulat Français

(1889 - 1894)

French consulate was moved repeatedly cause by a fire, and etc.

年代 Period	所在地	Location
1859 ～ 1862	神奈川宿慶運寺	Keiunji Temple, Kanagawa-Shuku
1862 ～ 1864	駒形町中横丁	Komagata-Machi Naka-Yokocho
1865 ～ 1866	居留地 31 番	Lot #31
1866 ～ 1866	運上所向側	front of the business tax office
1866 ～ 1875	本町 5 丁目	Honcho 5 chome
1875 ～ 1885	居留地 74 番	Lot #74
1886 ～ 1888	居留地 24 番	Lot #24
1889 ～ 1895	居留地 84 番	Lot #84
1896 ～ 1899	山手居留地 185 番	Yamate lot #185

居留地 84 番から発信されたカバー [fig.22]。本カバーが出状された 1894 年当時の領事はクロブコウスキィでした。

fig.22 is a cover sent from lot #84. In 1894, when the cover was sent, the consul was A. Klobukowski.

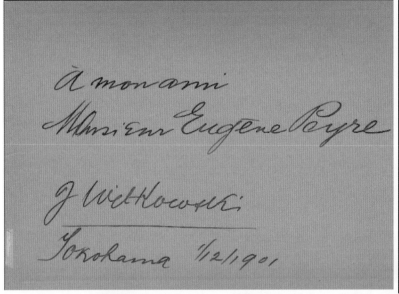

fig.22 フランス領事クロブコウスキィ (領事館 84 番) French Consul A. Klobkowski (Consulate, Lot #84)

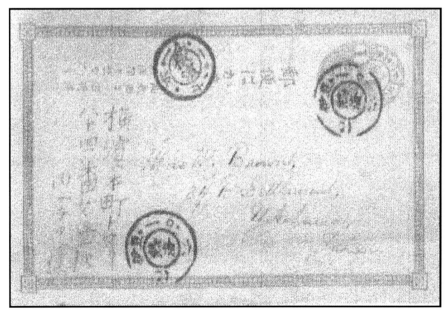

fig.21 横浜英和女学校ブラウン校長宛 1886 年 10 月 6 日 於熱海 カティ差立
To Director of Yokohama Eiwa Girl's School Miss Brown, Sent by E.C. Cathy in Atami, on October 6, 1886

なお、参考に居留地 74 番時代に領事館発信でフランス・セット（Cette。現在は Sète と綴る）宛の未納印付カバーを紹介します [fig.23]。この未納印付きカバーはこの 1 通だけが知られています。

For your information, here I show a cover with an unpaid postage mark, which was sent from the consulate to Cette (present Sète) when the consulate was on lot #74 [fig.23]. Only one cover with unpaid postmark has been recorded.

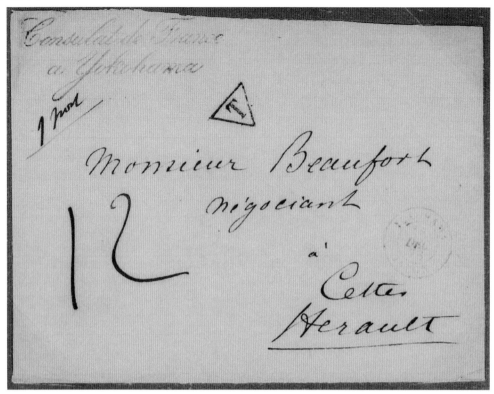

fig.23 フランス領事館 (74 番) 差立セット (現 Sète) 宛・未納扱
Sent from French Consulate to Cette (present Sète), Unpaid Postmark

fig.24 日本郵便切手総代理店 (85 番) 宛パリ差立葉書 Postcard from Paris to the General Agent of Japanese Stamps (Lot #85)

3-4 居留地 85 番の専有者の変遷

1882 年 8 月 12 日付 The Japan Weekly Mail 掲載：

1882 年 8 月 8 日未明、83 番のホッジス夫人方から出火、消火活動する間もなく全焼した。消火機器の到着後も給水が遅れ迅速かつ的確な消防活動が出来なかった。そのため炎は 83 番を焼き尽くし、ペィル兄弟ホテルに燃え移り、ホッジス夫人宅ともども焼失させた。隣接地の日本人居住区の一画も火と放水で被害を受けた。また、焼け落ちた柱などが燻っており、炭化した木材が 8 時頃まで時折炎を上げていた。2 時頃に出火、4 時までに 3 棟の床、屋根、壁の大部分が灰燼と化したほど火勢は強かった。犠牲者が出なかったのが何よりも幸いだった。ホッジス夫人は家具も家財もすべて焼失したが、ペィル兄弟は搬出しやすい高価な家具の一部を持ち出せた。ホッジス夫人宅のスイス人の使用人が頭と腕に重傷を負った。

3-4 Owners of Lot #85

The Japan Weekly Mail dated August 12, 1882, said:

Before the dawn of August 8, 1882, a fire broke out at lot #83, Ms. Hodges' residence, and it was destroyed without time for firefighting. Even the equipment arrived, but without an immediate supply of water. It impeded rapid and appropriate firefighting. Thus, a fire burned up the whole area of lot #83, spread to Peyre Brothers' Hotel to be burned up as Ms. Hodges' residence. The fire and water partly damaged the neighboring Japanese residential area. The burned pillars and other things were smoking, and some charred wood sometimes burst into the frame until around 8:00 a.m. A fire broke out around 2:00, and the flames were so strong that they burned up most of the floor, roof, and wall of three buildings by 4:00. It was a great relief that there was no victim. Although Ms. Hodges lost all the furniture and other possessions, the Peyre brothers rescued some luxury furniture easy to be taken out. A Swiss employee in the Hodges was seriously injured in the head and arm.

j. ペィル兄弟洋菓子店

（専有期間：1882 ～ 1899 年）

ペィル兄弟は 84 番のホテルが全焼したためホテル業から撤退し、隣接する 85 番に移転して洋菓子店を 12 月 4 日に新装開店しました。

j. Peyre Brothers' Patissiers and Confectioners

(1882 - 1899)

The Peyre brothers gave up the hotel business because of burning up the hotel on lot #84 and newly opened the confectioner on lot #85 on December 4.

k. 日本郵便切手総代理店

（専有期間：1882 ～ 1899 年）

ペィル兄弟次男マティウ＝ウジェンヌは 1887 年頃に「日本郵便切手総代理店」を創設し、洋菓子製造販売、食品。酒類の輸入販売やレストランと喫茶店経営の傍ら、日本切手をフランスを中心として欧州各国への輸出を手掛けていました。パリ差立 85 番の日本郵便切手総代理店宛カバー [fig.24] を紹介します。

k. General Agent of Japanese Stamps / Agence Générale des Timbres-Poste du Japon

(1882 - 1899)

The 2nd son of the Peyres, Mathieu Eugéne, established the "General Agent of Japanese Stamps" in around 1887 and exported Japanese stamps to Europe, mainly France, while he manufactured and sold the sweets, sold foods and alcohol, and managed the café-restaurant. Here I show a cover from Paris to the General Agent of Japanese Stamps on lot #85 [fig.24]

おわりに

1868 年から 1899 年までの横浜外国人居留地所在のすべての外国人の借地権や居住権の専有期間を一覧表に纏めつつあります。また、居留地 1 番から各地番発着の郵便物の入手に努めていますが道は遠い。

今回、スタンペディア創業者の吉田 敬さんから寄稿要請をいただき、従来各誌に掲載いただいた拙稿を参考にして本稿を作成しました。在横浜フランス郵便局はじめ横浜外国人居留地発着郵便物に興味をもっていただければ望外の幸せです。なお、居留地 80 番、84 番、85 番の専有者の移り変わりを一覧表に纏め、付録として添付しました。

Conclusion

I'm organizing the data of the settlement of all the foreigners in Yokohama. It will describe who had the leasehold or the right of residence between 1868 and 1899. Also, I'm collecting the mail of every lot from #1, but it is still a long way to complete.

I have an opportunity to make this article, referring to my articles published in some magazines before, to contribute to Stampedia. I'll appreciate it if you become interested in the French PO in Yokohama and the mail from/to the foreign settlement in Yokohama. The list of the history of possessors of lot #80, #84, and #85 is attached as an appendix.

Place	1868	1869	1870	1871	1872	1873	1874	1875
No.80	Berger & Co	Berger & Co	Berger & Co	Berger & Co				W.P.Brown
	Alfled Culty	Alfled Culty	Alfled Culty					Mr and Mrs Camargo
	Orney	Orney	Orney	Orney	Orney	Orney	Orney	P. Barruca
	The Sacred Heart of Jesus	The Sacred Heart of Jesus	Catholic Mission	The Sacred heart of Jesus	Catholic Church	Catholic Church	Catholic Church	Catholic Church
			E.Schwartz	E.Schwartz	E.Schwartz	E.Schwartz	E.Schwartz	E.Schwartz
	Japan Gazette Office	Japan Gazette Office	Unoccupied		J. Esdale			
					Cock Eye	Cock Eye	Cock Eye	Cock Eye
								General store and Frebch Bakery
No.84	Japan Gazette Office	Japan Gazette Office	Japan Gazette Office	Japan Gazette Office	Japan Gazette Office	Japan Gazette Office		
			J.M.. Jacquemot	J.M. Jacquemot	J.M. Jacquemot	Oriental Hotel	Oriental Hotel	Oriental Hotel
No.85	Kirby & Co.	Kirby & Co.	Kirby & Co.	Kirby & Co.	Kirby & Co.	Kirby & Co.	Japan Gazette Office	Japan Gazette Office
						F.W.Marks	Mrs.E.A. Vincent	Mrs.E.A. Vincent
								M.T. Aratoon
								J.Mayers
								Dr.J.J.R. Dalliston
								Saml. Parry

Place	1884	1885	1886	1887	1888	1889	1890	1891
No.80		C.H. Geffeney	C.H. Geffeney	C.H. Geffeney	C.H. Geffeney	C.H. Geffeney	C.H. Geffeney	C.H. Geffeney
			German Consulate		Grands Magasins du Printemps, Paris			Kuhn & Co.
					M.M. Rahimkhan	M.M. Rahimkhan	M.M. Rahimkhan	
	Catholic Church	Catholic Church	Catholic Church	Catholic Church	Catholic Church	Catholic Church	Catholic Church	Catholic Church
	Sargent Falsari & Co.	Sargent Falsari & Co.					Daibutsu	
				L.Harlow	L.Harlow	L.Harlow	L.Harlow	L.Harlow
	Cock Eye	Cock Eye	Cock Eye	Cock Eye	Cock Eye	Cock Eye	Cock Eye	Cock Eye
	Dominique Campana	Dominique Campana					T.Batchelor	T.Batchelor
	Mrs. Centurioni	Mrs. Centurioni	Mrs. Centurioni	Mrs. Centurioni	Mrs. Centurioni	Mrs. Centurioni	Mrs. Centurioni	Mrs. Centurioni
No.84	Connercial & Family Hotel	Connercial & Family Hotel	Lampert's Family Hotel	Methodist Protestant Misson School	Methodist Protestant Misson School	Methodist Protestant Misson School	French Consulate	French Consulate
No.85	Peyre Frères	Peyre Frères	Peyre Frères	Peyre Frères	Peyre Frères	Peyre Frères	Peyre Frères	Peyre Frères
	Mrs.E.A. Vincent	Mrs.E.A. Vincent	Mrs.E.A. Vincent	Mrs.E.A. Vincent	Mrs.E.A. Vincent	Mrs.E.A. Vincent	Mrs.E.A. Vincent	Mrs.E.A. Vincent
	Capt.A.F. Christensen	Capt.A.F. Christensen			G.Adet	G.Adet		J.Curnow & Co. Godown
	B.Roth	B.Roth			G. Campredom	G. Campredom		
			G. Naterman	G. Naterman	H.A. Vincent	H.A. Vincent	H.A. Vincent	H.A. Vincent
				A.G. Green	A.G. Green	W.Graham	W.Graham	W.Graham
	Capt.Jphn Steadman	Capt.Jphn Steadman				Capt.H.J. Carrew	Capt.H.J. Carrew	
				O.Letourneur		O.Letourneur	O. Jouvet	

横浜外国人居留地80番、84番、85番の専有者一覧表

Place	1876	1877	1878	1879	1880	1881	1882	1883
No.80	Peyre Frères	Peyre Frères	Peyre Frères		Baron Stillfried	Baron Stillfried	Baron Stillfried	Baron Stillfried
		A.F. Negre	A.F. Negre	A.F. Negre		H.Elfen		G. Zancolo & Co.
	P. Barruca	P. Barruca	F.Retz	F.Retz	F.Retz	H.Juery		T.R. Green
	Catholic Church	Catholic Church	Catholic Church	Catholic Church	Catholic Church	Catholic Church	Catholic Church	Catholic Church
	General Store French Bakery	A.Damiot & Co.		P.Marmande	Sargent Falsari &Co.	Sargent Falsari &Co.	Sargent Falsari &Co.	Sargent Falsari &Co.
	Patisserie Charcuterie Parisienne	Boulangerie Parisienne	Boulangerie Parisienne	Boulangerie Parisienne				Quick & Co.
	Cock Eye	Cock Eye	Cock Eye	Cock Eye	Cock Eye	Cock Eye	Cock Eye	Cock Eye
	General store and Frebch Bakery	B. Roth		Dominique Campana	Dominique Campana	Dominique Campana	Dominique Campana	Dominique Campana
		Mme.H. Giaretto		Mrs. Centurioni	Mrs. Centurioni	Mrs. Centurioni	Mrs. Centurioni	Mrs. Centurioni
No.84		Paul Sarda J.Lescasse	J.Lescasse	J.Lescasse		R.Gebauer		Temperance & Family Hotel
	Oriental Hotel	Oriental Hotel	Oriental Hotel	Peyre Frères Hotel	Peyre Frères Hotel	Peyre Frères Hotel	Peyre Frères Hotel	Peyre Frères Hotel
No.85	Japan Gazette Office	Japan Gazette Office	Japan Gazette Office	Jno.W.Hall	G.M. dos Remedios	Capt.A.F. Christensen	Capt.A.F. Christensen	Capt.A.F. Christensen
	Mrs.E.A. Vincent	Mrs.E.A. Vincent	Mrs.E.A. Vincent	Mrs.E.A. Vincent	Mrs.E.A. Vincent	Mrs.E.A. Vincent	Mrs.E.A. Vincent	Mrs.E.A. Vincent
		C.J.Strome		Dr.Wm. Anderson	Dr.Wm. Anderson		R.Gebauer	
	N. Stibolt	Capt.A.F. Christensen	Capt.A.F. Christensen	L. Davis	L. Davis		J.H.Curtis	
	Dr.Goertz			E.A. Bird			Capt.Jphn Steadman	Capt.Jphn Steadman
	Saml. Parry			H.Juery	H.Juery			H.Juery

Place	1892	1893	1894	1895	1896	1897	1898	1899
No.80			A.W.Cabeldu & Co.	A.W.Cabeldu & Co.	G.Schneider P.Launay	G.Schneider P.Launay	G.Schneider P.Launay	G.Schneider P.Launay
	The Yokohama Photographic Co.	Depot of Swiss Watches	Depot of Swiss Watches	Depot of Swiss Watches	Depot of Swiss Watches	Depot of Swiss Watches	Depot of Swiss Watches	Depot of Swiss Watches
	H.Campbell		S.Komor & Co.	Business Printing Office	Business Printing Office	Sung Wo K. Usui	Sung Wo K. Usui	Sung Wo K. Usui
	Catholic Church	Catholic Church	Catholic Church	Catholic Church	Catholic Church	Catholic Church	Catholic Church	Catholic Church
		India and Japan Co.	India and Japan Co.	India and Japan Co.	Arthur T. Watson Mcilraith, Crombie & Co.	Arthur T. Watson Ung-Ki	C.E. Miller J.Alfred Hart Hep Cheong	India and Japan Co.
	L.Harlow	F. Littlewood	King Tai	King Tai D.Asai	D.Asai C.E. Miller	D.Asai C.E. Miller	P.Mourier B.Monteggea	B.Monteggea Hep Cheong
	Cock Eye	Cock Eye	Cock Eye	Cock Eye	Cock Eye	Cock Eye	Cock Eye S. Kihara	Cock Eye S. Kihara
	T.Batchelor	T.Batchelor	T.Batchelor	T.Batchelor	T.Batchelor	T.Batchelor	T.Batchelor Yee Woo	T.Batchelor Yee Woo
	Mrs. Centurioni	Mrs. Centurioni	Mrs. Centurioni	Mrs. Centurioni	Mrs. Centurioni	Mrs. Centurioni	Mrs. Centurioni	Mrs. Centurioni
No.84	French Consulate	French Consulate	French Consulate	French Consulate			A.P. Sadra	A.P. Sadra
				G.Goudareau F.Sarzin				
No.85	Peyre Frères	Peyre Frères	Peyre Frères	Peyre Frères	Peyre Frères	Peyre Frères	Peyre Frères	Peyre Frères
	Mrs.E.A. Vincent	Mrs.E.A. Vincent	Mrs.E.A. Vincent	Vincent, Bird & Co.	Vincent, Bird & Co.	Vincent, Bird & Co.	Vincent, Bird & Co.	Vincent, Bird & Co.
	J.Curnow & Co. Godown	J.Curnow & Co. Godown	J.Curnow & Co. Godown	J.Curnow & Co. Godown	J.Curnow & Co. Godown	J.Curnow & Co. Godown	J.Curnow & Co. Godown	J.Curnow & Co. Godown
					E.Adet	E.Adet	E.Adet	E.Adet
	H.A. Vincent	H.A. Vincent	H.A. Vincent					
	W.Graham	W.Graham						
			Capt.W. Thompson	Capt.W. Thompson				

List of Possessors of lot #80, #84, and #85 of the French Settlement in Yokohama

References

01. The Japan Gazette, Hong List & Directory 1868 ～ 1911

02. The Japan Weekly Mail, Aug.12 ～ 26, 1882

03. 横浜開港資料館編「図説　横浜外国人居留地」（1998, 有隣堂）
    Foreign Settlement in YOKOHAMA (YOKOHAMA ARCHIVES OF HISTORY, 1998, YURINDO)

04. 神奈川県立博物館編「横浜銅版画」（1982, 有隣堂）
    Copper Engraving of YOKOHAMA(KANAGAWA PREFECRUAL MUSEUM OF CULTUAL HISTORY,1982, YURINDO)

05. 松本 純一「フランス横浜郵便局」、（2008, 星雲社）
    French Post Office in YOKOHAMA, (Jun-ichi MATSUMOTO, 2008, SEIUNSHA)

06. Jun Ichi Matsumoto「A History of the French Post Office of Yokohama」(2012, JPS)

07. 松本　純一「デグロン君カバー研究史」（2010, 日本郵趣協会）
    A history of Study of DEGURON-KUN covers, (Jun-ichi MATSUMOTO, 2010, JPS)

08. 鳥居　民「横浜山手・日本にあった外国」（2009, 草思社）
    Foreign Country inside Japan, (Tami TORII, 2009, Soshisya)

09. 切手研究会「切手研究」、2002 年 7 月 413~415 合併号、2004 年 3 月 422~423 合併号、2005 年 6 月
    426~428 合併号、2007 年 7 月 435~436 合併号、2010 年 3 月 445~446 合併号
    Philatelic Study vol. 413-415, 422-423,426-428,435-436,445-446 (2002-2010, Philatelic Study Group)

The Author：KOBAYASHI Akira

Collecting：Early Franco-Japanese Postal History

Awards：

  Early Franco-Japanese Postal History　　　LV (Paris 1999, Taipei 2017)

Membership & Qualification and etc.：

  Royal Philatelic Society, London since 2010

  Society for Promoting Philately (Japan)

太平洋開戦直前 〜主に 1941 年における〜 外国郵便
International Mails Just Before the Break of the Pacific War, Mainly in 1941
岩崎朋之 IWASAKI Tomoyuki

本稿の背景と主題

1939 年 9 月、ドイツのポーランド侵攻を皮切りに第二次世界大戦が勃発。同月にはイギリスおよび、フランスがドイツに宣戦布告を行いました。そして、ソ連もナチスによる支配地域拡大への対応や領土的野心から、フィンランドへの侵攻を開始するとともにバルト三国を併合。さらにドイツは、ノルウェー、ベネルクス三国、フランスなどに次々と進攻し、欧州全域にその戦線を拡大していきました。1941 年 6 月には、ドイツがソ連へ侵攻し、独ソ戦の火蓋が切られました。

そして、12 月 8 日、日本はマレー半島を攻撃するとともに、真珠湾も攻撃。日本はイギリスとアメリカ、オランダなどの連合国に対して開戦し、ドイツやイタリアもアメリカに宣戦布告、戦争は全世界的に拡大していくこととなりました。

Background and Theme

In September 1939, WWII broke out by the German invasion of Poland. In the same month, GB and France declared war against Germany. The USSR also started invading Finland and amfalgamated the three Baltic States. They did it against the Nazis who expanded their territory, and, as well as Germany, by their territorial ambition. Additionally, Germany invaded a series of countries such as Norway, Benelux, and France. The battlefield expanded to the whole of Europe. In June 1941, Germany advanced to the USSR; the German-Soviet War broke out.

Then, on December 8, Japan attacked Pearl Harbour as well as the Malay Peninsula. Japan declared war against the Allies of WWII, such as GB, the U.S., and the Netherlands. Also, did Germany and Italy against the U.S. The war expanded over the world.

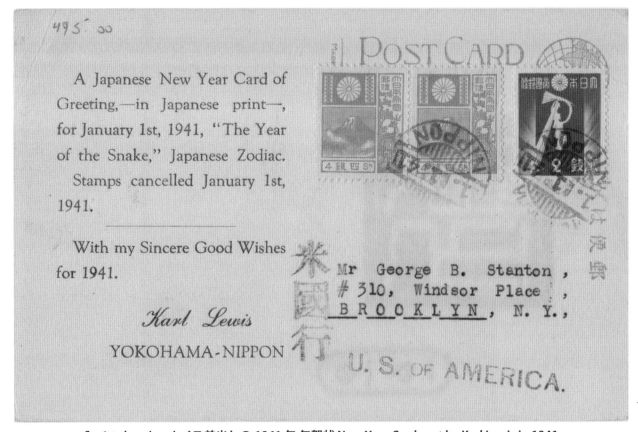

fig.1：カール・ルイス差出しの 1941 年 年賀状 New Year Card sent by Karl Lewis in 1941

こういった背景の中、日本はもとより、グローバルにおいて、外国郵便は極めて多大な影響を受けることとなります。交通網への影響はもちろん、宛先国、発信国、また通過国それぞれの関係性によって、郵便物の送達、通過が困難になりました。また、各国で郵便物に対して検閲が行われ、遅延や押収も頻繁に見られる状況となりました。そして、そういったケースを回避し、なんとか送達ルートを確保しようと、各国行政は郵便物を含めた物資の往来を確保するため、多大な労力を強いられることとなります。

とりわけ 1941 年に入ると、欧州方面では独ソ戦の勃発、そして、太平洋方面では日米の緊張が極限まで高まり、日本をとりまく外国郵便の状況が大きく、目まぐるしく変化していくこととなっていきます。本稿は、太平洋戦争直前期、特に 1941 年に入って以降、太平洋開戦までにおける外国郵便について、さまざまな実例を通してその状況を解説していきたいと思います。

図 1 は 1941 年における、画家、写真家であるカール・ルイス差出しの外国宛年賀状です。多くの美しいカバーを残したカール・ルイスも、1941 年には検挙され、軟禁状態におかれるなど苦難もあり、長年に渡り行われたカバー制作も本年発行の台湾の国立公園を最後に終了、翌年没しました。

That affected gravely not only to the Japanese but global international mail system. The influence on transportation, of course, and the change of relationship with sender and transit country made it difficult to send or transit mails. Each country censored mails, and frequently mail delivery delayed, or some mails were seized. Each government had to work hard to maintain the route avoiding these cases for the secure delivery of goods, including mails.

Especially in 1941, since the German-Soviet War broke out in Europe and the tension between the U.S. and Japan became extreme in the Pacific, the international mail situation around Japan changed significantly and rapidly. I describe in this article the situation of the international mail just before the Pacific War, especially in 1941, showing various examples.

fig.1 is an international New Year Card sent in 1941 by Karl Lewis, who was a painter and photographer. Karl Lewis, who left many beautiful covers, also had a hard time in 1941 by being arrested and confined. And the last work of his many years of creating covers was that of a national park in Taiwan of the same year, and he was dead the next year.

fig.1：カール・ルイス差出しの 1941 年 年賀状（裏面）New Year Card sent by Karl Lewis in 1941, the back

北米方面

　多彩な客船で賑わいを見せた北米航路も、1940年代以降、国際的な緊張の高まりにより次第に萎縮を強めていきます。

　図2は、定期便船停止前末期となる1941年4月に氷川丸船内より差し出された姫路城連合葉書です。北米航路の船内からソ連領内に宛てられた、このような使用例はほとんど見かけないものです。

　この宛先のオレシコ（Olesko）は、現在のウクライナ・リヴィウ州の都市であり、第一次大戦後のウクライナ独立時にはその首都が置かれていました。しかし、その後、ポーランドに占領されたのち、第二次大戦でドイツの侵攻を受け、独ソの秘密協定により、この地域はソ連に割譲されました。

　この使用例はこれ以降に宛てられたもののため、宛先国はソ連となっています。文面は表裏に渡って、びっしりと書かれていますが、非常に馴染みの薄い文字で書かれています。これは中・東欧のユダヤ人によって使用されていたイディッシュ語の文面と思われます。

　恐らくこの通信は、欧州から逃れたユダヤ人がシベリア鉄道経由で日本に至り、その後、最終目的地の米国に向かう船内にて、故郷に宛てたものと考えられます。なお、1941年6月の独ソ戦でこの地域もドイツの侵攻を受けることとなります。そして、直後にリヴィウポグロムというユダヤ人に対する大規模な迫害、虐殺事件が発生することとなります。名宛人が不幸にもこの事件に巻き込まれてしまった可能性も考えられます。

North America

The North American sea route, which showed prosperity with various passenger boats, was getting declined in the 1940s by high international tension.

fig.2 is a U.P.U. postalcard of Himeji Castle design, which was sent from a liner HIKAWA MARU in April 1941, just before the suspension of regular liners. It was an infrequent usage that was posted on a ship and carried by the North American route to the USSR.

The address, Olesko, was a city in Lviv, current Ukraine, which was the capital of the country after WWI when Ukraine became independent. However, Poland occupied there the next, and Germany invaded during WWII. The secret agreement between Germany and the USSR decided to cede this region to the USSR.

This usage was sent after the agreement, so the address was the USSR. The postalcard was full of text but an uncommon language for Japanese people. It was possibly Yiddish, a Jewish language used in central and eastern Europe.

Maybe who sent this mail was a Jew, who escaped from Europe to Japan using the Trans-Siberian Railways. The sender wrote to his/her family traveling to the U.S., the final destination. However, Germany invaded there during the German-Soviet War in June 1941. Moreover, just after then, the constant persecution and massacres of Jews called Lviv Pogroms took place. The addressee might have been a victim of this incident.

fig.2：ソ連領時代のウクライナ・オレシコ宛葉書。氷川丸船内差出し
Postalcard to Olesko (Ukraine) under the Soviet occupation, sent from a liner Hikawa Maru

日中戦争や仏印進駐により、日米関係が決定的に悪化、日本に対する輸出入の禁止もあいまり、北米における各社定期便船は停止されていくこととなりました。1941年7月、日本船のパナマ運河通航が禁止となり、さらに、日本の在外資産は凍結されるなど、日本に対する厳しい制裁措置が発動されていくこととなります。この時期に日本から米国に宛てられた連続便があるので、図3をご覧ください。

図3はこの時期における米国宛の同一差し出し、同一受取人の連続便です。太平洋は船舶にて逓送され、サンフランシスコからは東海岸へ向け、航空路にて逓送されたものです。7月10日の東京消印のものは、7月10日出港の龍田丸の指定、7月29日のものは8月7日出港の新田丸の指定がされています。そして、それ以降のものについては定期便船が停止されたため、便船指定がなくなります。外国郵便が非常に減少していくこの時期に、このような形で連続して差し出された使用例が残されていることは稀であり、これらは非常に貴重な資料と言えるでしょう。

The Sino-Japanese War and the Japanese invasion of French Indochina worsen the relationship between the U.S. and Japan decisively. Also, the U.S. prohibited Japan from importing and exporting goods. Under the circumstances, the private companies' regular liners in North America became suspended. In July 1941, the U.S. invoked strict sanctions against Japan: they prohibited Japanese ships from passing the Panama Canal and froze Japanese assets overseas. I show in fig.3 the correspondence from Japan to the U.S. in the period.

fig.3 is a series of correspondence to the U.S. between the same sender and addressee. Ships carried them over the Pacific Ocean and airlines from San Francisco to the east coast. The letter with a Tokyo cancellation of July 10 designated TATSUTA MARU left on the same day, and that of July 29, NITTA MARU, which was scheduled on August 7. The last one, because of the suspension of regular liners, didn't designate the liner. A series of correspondence of international mail during the international-mail-declining period was uncommon. So, these are valuable documents.

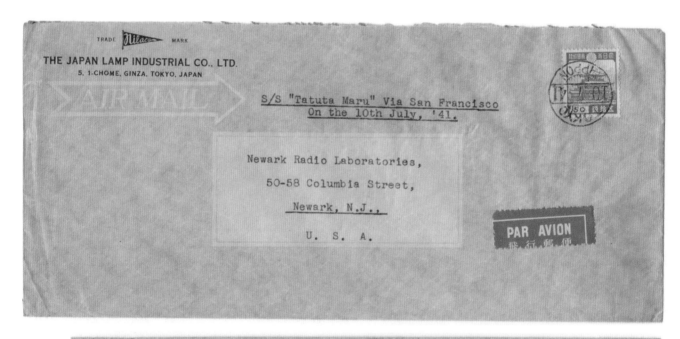

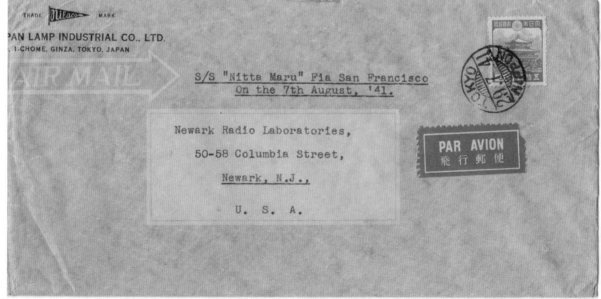

fig.3：定期便船停止期の米国宛連続便 Correspondence to the U.S. during the suspension of regular liners

この龍田丸は、7月10日に横浜出港後、7月23日にサンフランシスコ沖に到着しましたが、在米日本資産凍結通告により、船体や積荷が没収される危険がありました。それを回避するため、両国政府における交渉が行われたため入港ができず、7月30日になってようやくサンフランシスコ入港が叶いました。しかしながら、それまで船舶抑留はしないとしていた米国政府が、8月2日以降はそれを保証しないと表明したことから、急遽行われた帰港命令により十分な準備が行われずに出港をしたため、船中で大量の食中毒が発生し、100名以上が発症、9名が死亡するという惨事にみまわれました。

なお、この龍田丸の一便あとの7月18日横浜出港の浅間丸は、サンフランシスコへ到着できず、途上で反転し横浜へそのまま帰港しています。また、8月7日出港予定の新田丸の方は、本航海は中止となったと見られ、その後、1941年9月付で日本海軍に徴傭されています。

この後、日本政府は米国と交渉を行い、在留邦人帰国のための引揚船を運行することとなります。北米方面では、10月15日の龍田丸、10月20日の氷川丸などが運行されました。これらはそれぞれ、数百名の引揚者を日本へ帰国させました。

Above mentioned TATSUTA MARU left Yokohama Port on July 10 and arrived offshore San Francisco on July 23. However, the U.S. and Japanese governments had to negotiate for avoiding the seizure of the ship and goods by a freeze of Japanese assets in the U.S., and the ship couldn't enter the port immediately. Finally, on July 30, it arrived in San Francisco. Still, the ship got into trouble. The U.S. government, which said not to intern the ship, declared not to guarantee that from August 2. So, the ship had to leave by a sudden return-to-port order without sufficient preparation, and many food poising cases broke out on board. The disaster concluded with a hundred or more patients and nine deaths.

ASAMA MARU, the next ship of TATSUTA MARU, departed Yokohama on July 18 but couldn't arrive at San Francisco and returned Yokohama. And NITTA MARU, which planned to leave on August 7, was possibly suspended the departure, and in September 1941, it was requisitioned by the Japanese navy.

After then, the Japanese government and the U.S. agreed on operating a repatriation ship for Japanese residents in the U.S. TATSUTA MARU departed on October 15 and HIKAWA MARU on October 20 did that service in North America. Each ship took some hundred people to Japan.

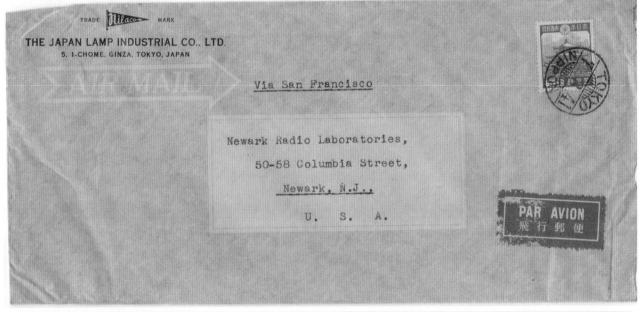

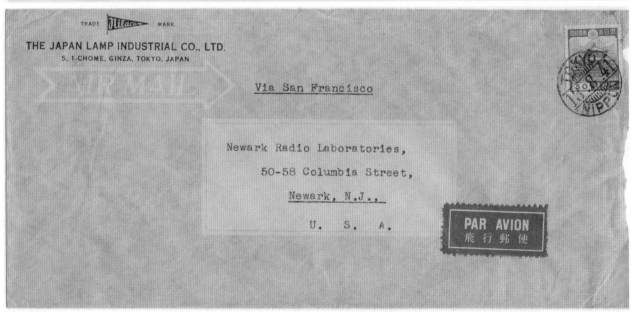

北米方面の船便による郵便については、引揚船が運行された10月下旬あたりが確認される最終期の使用例となります。これらが、図4,5となります。図4は、ホノルルの在留邦人から1941年10月15日消印で神戸に宛てられた使用例です。なお、本例では、上海経由が記載されており、外国船での逓送が指定されたものです。また、図5は、カナダ宛の使用例であり、10月18日の軽井沢消印が押捺されています。末期の使用例となりますが、カナダでの検閲印も押されており、実際に現地まで到着しています。

The latest usages of the mail to/from North America were at the end of October when the repatriation ships operated. These are shown in fig.4 and 5. fig.4 was sent to Kobe by a Japanese resident in Honolulu, canceled on October 15, 1941. This letter said "via Shanghai" and designated a foreign ship. fig.5 is the usage to Canada with a Karuizawa cancellation of October 18, which was one of the latest usages. It was censored in Canada and able to arrive at the addressee.

fig.4：1941年10月のハワイ発日本宛書状 Letter from Hawaii to Japan in October 1941

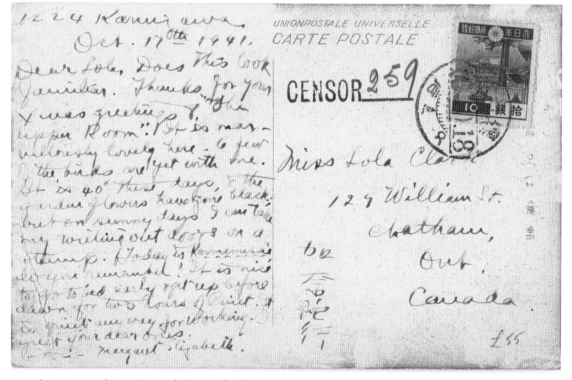

fig.5：1941年10月の日本発カナダ宛葉書 Postalcard from Japan to Canada in October 1941

さらに、興味深い使用例がありますので、図6をご覧ください。ニューヨーク 1941 年 10 月 17 日の消印がある本使用例は、同地在留の邦人から発信されたものであり、フィラデルフィア見物といった気軽な文章とともに、日本の「政変」という記載もあり、緊迫した情勢が当地にも伝達されているのが伺えます。1941 年 10 月 16 日に対米交渉のもつれなどにより、第 3 次近衛内閣が総辞職していますので、これを指しているのでしょう。

I show here one more fascinating usage in fig.6. It is a postcard with a New York cancellation of October 17, 1941. It was sent by a Japanese resident, writing about happy sightseeing in Philadelphia, also about serious "change of Government" in Japan. A tense situation seemed to be known there. The possible "change of Government" then was the 3rd Konoe cabinet's resignation on October 16, 1941, which the complications in the negotiations with the U.S. triggered.

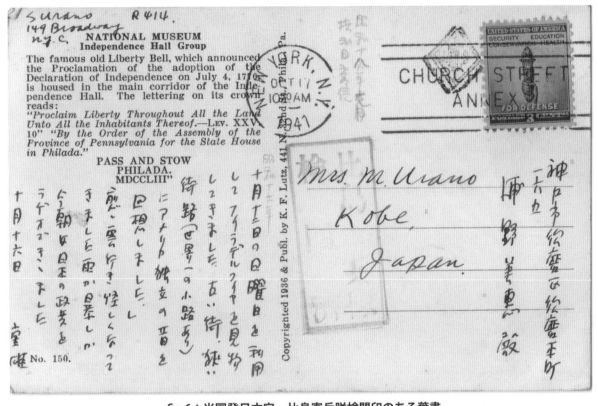

fig.6：米国発日本宛、比島憲兵隊検閲印のある葉書
Postcard from the U.S. to Japan, the postcard with the censor mark of Philippine military-police unit

この使用例は、先に記述した 10 月 15 日日本出港の龍田丸に搭載される予定でした。しかしながら、情勢悪化により、郵便物の龍田丸搭載が拒否され、龍田丸は急ぎ帰郷することとなりました。これにより、搭載されるはずだった日本向け郵便物は、11 月 8 日出港の米国船、プレジデント・グラント号に搭載されることとなりました。このプレジデント・グラント号は極東方面に出港、12 月 4 日マニラに到着し、郵便物はマニラにて陸揚げされたのち、マニラ郵便局にて長く滞留されることとなったのです。

その後、日米開戦となりフィリピンを占領した日本軍は、マニラ郵便局に滞留されたこれらの郵便物を発見、それらに憲兵隊の検閲を実施したのち、1943 年初旬に配達されることとなりました。この葉書の表面には「昭和十八年一月拾弐日受信」との書込みがされています。この処理を受け送達された郵便物はかなりの数に上ると見られますが、現在ではわずかに確認されているに過ぎません。

This postcard was planned to be carried by TATSUTA MARU, which left Japan on October 15. However, the situation was worsening, and TATSUTA MARU negated to carry the mail and had to return immediately. Finally, a U.S. ship, PRESIDENT GRANT, which departed on November 8, carried the planned mail to Japan. This ship left for the Far East, arrived in Manila on December 4. The mail was discharged and kept at Manila P.O. for a while.

Then, the war between the U.S. and Japan broke out. When the Japanese troops occupied the Philippines, they found the mail suspended in Manila P.O. The military-police unit censored them, and finally, the mail reached the address at the beginning of 1943. This postcard has a handwritten description of "received on January 12, Showa 18 (1943)." Although possibly there was a lot of mail delivered by this step, we have found only some usages.

北米方面について、最後に航空路をご紹介したいと思います。米国の航空会社パンナムは、サンフランシスコからホノルルを経由し、フィリピンを経て香港に至る路線を、クリッパー飛行艇により開拓しました。これにより、日本 - 米国間の郵便物を香港経由で、太平洋を空路にて横断し逓送できることになりました。しかしながら、このクリッパー艇による航空郵便は料金が非常に高額になるため、料金も低廉でさほど日数も掛からない船舶による逓送が一般的であり、このクリッパー便を利用した日本発着の使用例はあまり多くは見られません。

図 7 は開戦前最末期のクリッパー便による使用例であり、1941 年 10 月 29 日付ボストンの消印が押捺され、朝鮮・京城に宛てられています。到着されたと思われる 11 月には、臨時郵便取締令がすでに発令されていたため、「朝鮮検閲済」の印が押され、逓信省による検閲シールが貼られています。これは朝鮮の京城局で行われた検閲と見られます。日本語による宛名の追記も書き込まれており、宛先には無事到着したと思われます。

I'd like to show airmils finally. Pan Am, the U.S. airway, created the route by clipper from San Francisco to Hong Kong via Honolulu and the Philippines. This route made it possible to deliver the mail between Japan and the U.S. via Hong Kong, over the Pacific Ocean by air. However, the postage rate was high, and the delivery by ship was still more common. It was because there were many ships operating services, and the delivery by ship didn't take so much time. So, there are not many usages of this clipper delivery to/from Japan.

fig.7 is the usage of clipper delivery almost the outbreak of the War. It was to Kyongsong, Korea, with a Boston cancellation of October 29, 1941. In the possible arrival month of November, the provisional mail control act was already issued. So, it has a mark of "censored in Korea" and a censorship sticker of the Ministry of Communications. I suppose the censorship was done at Kyougsong P.O. in Korea. This letter has a handwritten address in Japanese, and it must have been received safely by the addressee.

fig.7：末期の朝鮮宛クリッパーメール Clipper mail to Korea before the War

欧州方面

Europe

欧州方面へ逓送は、1941 年に入っても、シベリア鉄道経由便が変わらず機能しており、日本ならびに極東と欧州を短い日数で接続させていました。しかしながら、1941 年 6 月 22 日に勃発した独ソ戦により、シベリア鉄道を経由して欧州方面に郵便逓送を行うことが困難となり、長年極東と欧州を接続してきたこのルートもついに途絶することとなります。なお、このシベリア鉄道経由欧州方面ルートは、1942 年に、モスクワから、チフリス、エルゼラム、イスタンブールを経て、欧州方面に入る迂回ルートを用いることにより、再開をしています。

シベリア鉄道経由での欧州との接続が途絶して以降は、主に東廻りのルートで欧州方面に接続することになりま

The mail delivery to Europe didn't change and the Trans-Siberian Railways was still available at the beginning of 1941. It was made possible to connect Japan and the Far East to Europe in a short time. However, the outbreak of the German-Soviet War on June 22, 1941, made it hard to carry the mail to Europe by the Trans-Siberian Railways. Finally, this route, which connected the Far East and Europe for an extended period, was also suspended. The route for Europe via the Trans-Siberian Railways restarted in 1942 using a detour: Moscow - Tiflis - Erzurum - Istanbul -Europe.

After the suspension of connection to Europe via the Trans-Siberian Railways, a west-to-east route was mainly used for Europe. It went over the Pacific Ocean to Lisbon via Nassau and the Bermuda Islands. However, as described before,

した。これは、太平洋を横断し、ナッソーやバーミューダを経て、リスボンから陸上に入るというルートです。ただ、先に記述の通り、7月以降は太平洋航路の定期配船は停止し、10月頃を持ってほぼ逓送は困難となるため、この頃には東廻りでの欧州宛逓送もほぼ途絶することとなります。

この時期の面白い使用例として図8をご覧ください。1941年6月14日付の神戸消印のドイツ宛書状です。表面には乃木2銭切手のみ貼付しており、裏面に18銭分の切手が貼付されています。この書状、経由地が手書きで変更されており、加えて、若干不鮮明ですが、引受消印から幾分後の7月10日付の捨印が押してあります。これらは何を意味するのでしょうか？

the regular liner service over the Pacific was suspended from July, and the mail delivery became almost impossible in October. Then, the delivery to Europe via a west-to-east route almost stopped.

fig.8 is a remarkable usage of that time. It is a letter to Germany with a Kobe cancellation of June 14, 1941, which has only a 2sen Nogi stamp on the front and the stamps for 18 sen on the back. The handwriting corrected the transit place, and it has an additional postmark on July 10, some days after the acceptance mark. What does it mean?

fig.8：1941年6月のドイツ宛書状。シベリア経由からアメリカ-リスボン経由に変更したもの
To Germany in June 1941. Change the route via Siberia to via the U.S. and Lisbon.

差出人はドイツに宛てて、それまでと変わらず、シベリア経由指定でこの書状を差出します。そして神戸局で引き受けられますが、逓送途中の 6 月 22 日に独ソ戦が勃発、シベリア経由での逓送ができなくなります。神戸局では、このようなシベリア経由指定がされながら、そのルートでの逓送ができなくなった郵便物に対し、一枚の付箋を貼付し、差出人に返戻を行いました。その付箋には概ね次のような記載がありました。「本郵便物はシベリア経由指定で差し出されましたが、独ソ開戦のため返送されたため、差出人に返戻します。米国 - リスボン経由での送達は可能なので、それを希望の場合はその旨記載の上、付箋貼付のまま再差し出しをして下さい」。この付箋が貼付された使用例は何点か確認されています。

そして、この図 8 の使用例もこの付箋を貼付され、返戻されたのでしょう。よく見ると付箋を剥がした跡を確認することができます。この書状を受け取った差出人は、付箋記載の通り、シベリア経由表記を抹消のうえ、米国 - リスボン経由の旨、表記し直します。そして、差し出しされたこの書状は神戸局で再度引き受けられ、神戸局にて不要になった付箋を取り除いた上、再引受の表示印を余白に押捺、指定通り東廻りルートで逓送したものと見て間違いはないでしょう。

この時期の使用例として図 9-10 は、独ソ戦勃発によるシベリア経由便途絶以降、東廻りを意味するリスボン経由指定書込みのある欧州宛使用例です。リスボン経由指定が記載された使用例も案外少ないものです。図 9 は、1941 年 7 月に那須よりドイツに宛てられた葉書であり、リスボン経由指定が書き込まれています。また、図 10 は中国占領地となる青島よりドイツに宛てられた葉書であり、こちらは「マニラ - リスボン経由」の指定がされています。

The sender posted the letter as usual to Germany via Siberia. Kobe P.O. accepted it. But in the midway on June 22, the German-Soviet War broke out, and the delivery via Siberia became impossible. Kobe P.O. put a label on the mail which couldn't be delivered via Siberia and returned to the senders. The label said, "this mail has designated the route via Siberia, but we returned to the sender caused by the German-Soviet War. Since the U.S.-Lisbon route was available, if you want, you can post it again with this label indicating your wish." There are some usages with this label.

fig.8 is also a returned mail with the label. If you look carefully, you will find a trace of a label. The sender, who received it, struck through "via Siberia" and corrected "via the U.S.-Lisbon," as the label indicated. It is no doubt that the letter reposted was accepted at Kobe P.O. again, where the invalid label was removed, and a reacceptance mark was stamped in a margin, delivered by the designated west-to-east route.

I show usages in that period in fig.9 and 10. These are mail to Europe, which designated the route via Lisbon, meaning a west-to-east route, after the stoppage of the delivery via Siberia caused by the German-Soviet War. There are not so many usages with a designation of the route via Lisbon. fig.9 is a postcard sent from Nasu to Germany in July 1941, designated the route via Lisbon with handwriting. fig.10 was a postcard from Qingdao, the occupied territory in China, to Germany, and it designated the route "via Manila-Lisbon."

fig.9：リスボン経由指定のドイツ宛葉書 Postcard to Germany via Lisbon

fig.10：マニラ、リスボン経由指定の青島発葉書 Postcard from Qingdao designated the route via Manila and Lisbon

　欧州方面へのこの時期の末期の使用例として図11をご覧ください。蘆屋の 1941 年 10 月 19 日付の消印が押されています。東廻り便と思われますが、10 月下旬の欧州宛使用例は非常に少ないものです。

fig.11 is one of the latest usages to Europe. It had an Ashiya cancellation of October 19, 1941. I suppose it via a west-to-east route delivery, but anyway, the usages to Europe at the end of October are quite rare.

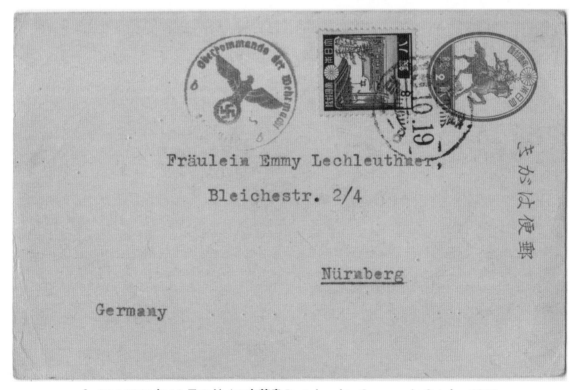

fig.11：1941 年 10 月のドイツ宛葉書 Postalcard to Germany in October 1941

さらに、独ソ戦勃発以降の、欧州方面から日本宛の使用例をお見せします。図12は、1941年8月21日イタリア・ジェノヴァ発、神戸宛使用例です。恐らく、欧州から出ることなく太平洋開戦を迎え送達不能となり、返戻となったと思われます。1942年2月の返戻の際の着印が見られます。

Now I show some usages from Europe to Japan after the outbreak of the German-Soviet War. fig.12 was sent from Genova, Italy, on August 21, 1941, to Kobe. It must have been returned because the Pacific War made it impossible to carry it out of Europe. It got an arrival mark in February 1942, when it returned.

fig.12：1941年8月のイタリア発日本宛書状 Letter from Italy to Japan in August 1941

また、図13は、1941年7月27日、フランス・ヴォクリューズから横浜に宛てられた使用例です。アメリカ経由記載がされており、リスボン - ニューヨークを経て日本に逓送されたと見られます。本例には、逓信省による検閲が実施されており、1941年10月の臨時郵便取締令以降に到着しています。逓信省の検閲封緘紙は用いられず、まだ有り合わせの用紙で封緘がされており、臨時郵便取締令初期、1941年10月上旬に到着したものと推定できます。

加えて、これに関連した使用例をご紹介します。1940年頃より、北米および中南米 - ドイツ間郵便物について、大西洋経由ではなく、太平洋を越えて、日本およびシベリア経由指定の郵便物が見られるようになります（図14）。本例はグアテマラ発信のドイツ宛葉書ですが、「日本 - シベリア経由」の記載がされています。次に図15ですが、図14とは逆の、ドイツから中米ホンジュラス宛の使用例となります。本例は1941年6月19日ベルリン発信であり、こちらにも、シベリア鉄道 - 日本経由指定のスタンプが押捺されています。しかしながら、折り悪く逓送途中の6月22日に独ソ戦が勃発し、シベリア経由での逓送が困難になります。そのため、「リスボン - ニューヨーク経由でのみ逓送可能」という内容の付箋が貼られたものです。

fig.13 was a letter sent from Vaucluse, France, on July 27, 1941, to Yokohama. The designated route was via the U.S. and possibly delivered to Japan via Lisbon and New York. It passed censorship of the Ministry of Communications and arrived after the announcement of the provisional mail control act in October 1941. It doesn't have an official censorship seal of the Ministry of Communications but a provisional one of ordinary paper. So, it is supposed to have arrived at the beginning of October 1941, just after the act took effect.

Some more related usages. Around 1940, they gave up to use the Atlantic route, and the Pacific route via Japan and Siberia worked as the mail-carrying route between America and Germany (fig.14). It is a postcard from Guatemala to Germany, with a description of "via Japan-Siberia." fig.15 is a usage of the reverse route of fig.14, from Germany to Honduras in Central America. The letter was sent from Berlin on June 19, 1941. It also has a stamped designation of "via the Trans-Siberian Railways and Japan." However, unfortunately on the way, on June 22, the German-Soviet War broke out, and it became hard to deliver via Siberia. Then, a label that said, "possible to deliver via Lisbon and New York," was pasted on the envelope.

fig.13：逓信省検閲のあるフラン
ス発日本宛書状
Letter from France to Japan
censored by the Ministry of
Communications

fig.14：グアテマラ発、
日本経由ドイツ宛葉書
Postcard from Guatemala
to Germany via Japan

fig.15：シベリア経由途絶の付箋があ
る、ドイツ発南米宛書状
Letter from Germany to South
America with a label saying the route
via Siberia was not available.

続いて航空路について見ていきたいと思います。欧州方面航空路は、シベリア経由や南アジア経由も停止しており、1941 年には欧州にはほとんど航空遞送が困難となっていました。この時期に可能であったのは、クリッパー艇での香港、米国経由など限られたルートのみとなっていました。そのような中で、1941 年 4 月 13 日、日ソ中立条約の締結によりシベリア経由からドイツに接続する航空郵便が、6 月 23 日より再開されることとなりました。しかしながら、6 月 22 日に独ソ開戦がなされるため、このルートの再開は幻となりました。図 16 は、この時期のシベリア経由航空便であり、イタリア・ジェノヴァの 1941 年 6 月 23 日消印で、関東州・大連に宛てられています。しかしながら、シベリア経由送達はこの差し立てと時を同じくして途絶するためこのルートによる航空送達は不可能となり、書込みなどから恐らく南アジア経由のルートを用いて大連まで送達がなされたものです。

Now I describe the air route. The air route to Europe was stopped even via Siberia and South Asia, and in 1941 the delivery to Europe became almost impossible. The available method in that period was by clipper, using some limited routes such as via Hong Kong or the U.S. In that situation, the Japan-Soviet Neutrality, which concluded on April 13, 1941, could restart the airmail system to Germany via Siberia on June 23. But on June 22, the German-Soviet War broke out, and the expected restart faded. fig.16 is an airmail via Siberia at that time. It has a Genova (Italy) cancellation of June 23, 1941, and was sent to Dalian, the Kwantung. But, as you know, the delivery via Siberia suspended just when the letter was sent, and it became impossible to use this route. By considering handwritten notes, it was possibly delivered via South Asia to Dalian.

fig.16：イタリア発大連宛航空書状 Airmail from Italy to Dalian

そして、欧州までの新たな航空路の確保への試みとして、南米経由での欧州宛航空ルートが 1941 年 8 月 1 日より開始されました。太平洋を船舶、またはクリッパー艇で越え、サンフランシスコ - ブラウンズビル - クリストバル - サンチャゴ - リオデジャネイロ、そして、ローマと繋ぎ、欧州に至るというルートになります。この南米経由ルートは 1941 年後半から太平洋を越えることが可能な時期までの、極めて限られた期間のみ可能なルートであったため、往復いずれについても、極めて僅かにしか使用例は発見されていません。

図 17 は、この南米経由で遞送された航空便です。1941 年 7 月 28 日のドイツ・ロットアッハ＝エーガーン、および 8 月 1 日のベルリン消印で抹消されています。そして、上部には、ローマ - リオデジャネイロ経由の表記がされ、南米経由航空路での遞送が指定されています。

Then, they tried having other air routes to Europe. On August 1, 1941, the route via South America opened. It was over the Pacific Ocean by ship or clipper, then, San Francisco-Brownsville-Cristobal-Santiago-Rio de Janeiro by ship, and finally to Rome, in Europe. Since the South American route was available in a quite limited period from the second half of 1941 to the time to be possible to pass the Pacific Ocean, extremely few usages have appeared.

fig.17 is an airmail via South America. It was canceled by Rottach-Egern (Germany) cancellation on July 28, 1941, and Berlin on August 1. It has a description of "via Rome-Rio de Janeiro" at the top to designate the route via South America.

また、図18は、スイス・チューリッヒ発横浜宛使用例であり、1941年8月13日のチューリッヒ消印が押されています。イタリア経由指定が記載され、これもローマ‐リオデジャネイロと繋ぐ南米経由航空ルートを念頭に差し出されたものと思われます。

　この時期になると、空路、海路ともに混乱を極めており、このいずれのカバーにおいても経由指定はあるものの中継印はなく、いかに欧州から日本まで到達したか、その詳細なルートは定かではありません。検閲シールなども頼りに、さらに調査が必要でしょう。

fig.18 is a usage from Zurich (Switzerland) to Yokohama, and it has a Zurich cancellation of August 13, 1941. It designated the route via Italy, so, the sender might have posted it expecting to carry it by the South American route, via Rome and Rio de Janeiro.

Both of air and surface routes were in confusion then, and both covers are with route instructions but without transit postmarks. So it's not clear how these covers were transported from Europe to Japan. We have to make a study more with knowledge of censor seals, and etc.

fig.17：ドイツ発南米経由日本宛航空書状 Airmail from Germany to Japan via South America

fig.18：スイス発イタリア（‐南米）経由日本宛書状 Letter from Switzerland to Japan via Italy (and South America)

南アジア方面

次は東南アジア方面を見てみたいと思います。図19は、蘭印駐在のビジネスマンより日本に宛てられた葉書です。1941年10月8日付、11月20日着となっており、かなり末期の使用例と言えるでしょう。

文面には、「九月十一日北野丸がバタビヤから日本向け出帆してから、彼此一ヶ月便船が途絶へ而もまだ近く便船は期待出来ず…」との記載があり、この方面において、便船がかなり途絶えがちとなっている状況が見て取れます。上海指定の書込みがあるのはこの辺りの事情と関連があるのでしょう。なお、この差出人は結局、太平洋戦争開戦まで蘭印に残り、その結果、敵国側に拘束され、オーストラリアの収容所に抑留されることとなりました。

また、図20は、1941年9月30日横浜発のスマトラ宛、姫路城連合葉書ですが、スマトラの敵国人収容所の抑留者に宛てられたものです。1940年のドイツによるオランダ侵攻により、蘭印方面のドイツ人は蘭印政府により拘束され、ジャワやスマトラの収容所へ抑留されることとなりました。この時期、日本からこのような収容所に宛てられた使用例が比較的多く見受けられます。この方面においては、1941年11月13日の日昌丸が最後の引揚船として記録されています。

航空路については、1940年から郵便遞送が開始された日泰航空が機能していました。東京から福岡、台北、広東、海口と経由するものであり、欧州諸国の航空路線も乗り入れるバンコクにルートを開くことにより、欧州やアフリカ方面への航空路の拡大も企図されたものです。

太平洋戦争開戦前末期の使用例として、次のfig.21, fig.22をご紹介します。図21は、1941年11月19日に、東京からハノイ宛てられた、日泰航空を用いた航空便です。現地には12月1日と、開戦直前に到着しています。臨時郵便取締令以降の使用例であり、「対照済」の記載と印鑑が押捺されています。臨時郵便取締令により、民間の郵便物は遞信省により検閲に付されましたが、大使館や政府関係の通信については、「対照済」と押捺または記

South Asia

Let's see Southeast Asia. fig.19 is a postcard sent by a businessman in Dutch Indies to Japan. It is a very late usage which was sent on October 8, 1941, and arrived on November 20.

The sender said, "on September 11, after the ship, KITANO MARU left Batavia to Japan, we have no mail ship in one month and cannot expect another ship soon." It shows the navigation service often stopped. Maybe the designation of the route "via Shanghai" related to this situation. The sender stayed in Dutch Indies until the outbreak of the Pacific War. He was held as a prisoner of war in a camp in Australia.

fig.20 is a U.P.U. postalcard of Himeji Castle design to Sumatra, sent from Yokohama on September 30, 1941. It was sent to a prisoner in an enemy camp in Sumatra. By the German invasion upon the Netherlands in 1940, German residents in Dutch Indies were held as a prisner of war in camps in Java or Sumatra by the Dutch Indiesn government. There are relatively many usages from Japan to those camps. The last repatriation ship recorded was NISSHO MARU departed on November 13, 1941.

As for air routes, Japan-Thai Airways, which started service in 1940, was still available. It used the route of Tokyo-Fukuoka-Taipei-Canton-Haikou. They intended to extend the route to Europe and Africa by opening the route to Bangkok, where European airlines extended the service.

fig.21 and fig.22 are the latest usages before the Pacific War. fig.21 was an airmail sent from Tokyo on November 19, 1941, by Japan-Thai Airways to Hanoi. It arrived on December 1, just before the war outbreak. Since it was after the announcement of the provisional mail control act, the handwriting of "contrasted" and the name seal were on the envelope. According to the act, the Ministry of Communications censored private mail, but they put the handwriting or handstamp of "contrasted" for the correspondence of the embassy or government. This usage was to the government-general in Hanoi.

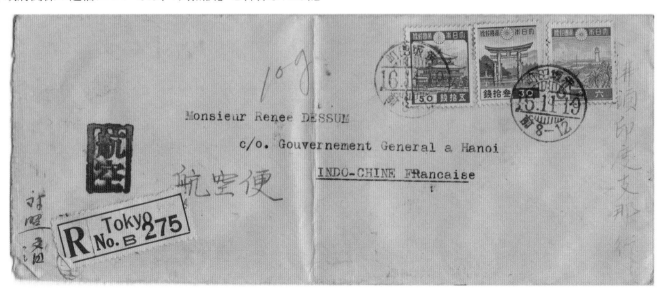

fig.21：ハノイ宛日泰航空経由航空書状 Airmail to Hanoi by Japan-Thai Airways

載を受けています。本例も、ハノイの総督府に宛てられてたものです。

　また、図22は、同じく日泰航空を利用したタイ宛郵便であり、1941年11月25日発信で、12月3日に到着しています。太平洋開戦前の外国郵便としては最末期のものと言えるでしょう。

fig.22 was sent to Thailand on November 25, 1941, via Japan-Thai Airways. It arrived on December 3. It must have been the latest international mail before the Pacific War.

fig.19：末期の蘭印発日本宛葉書
Postcard from the Netherlands Indies to Japan just before the War

fig.20：蘭印ドイツ人収容所宛姫路城連合葉書
UPU Postalcard of Himeji Castle design to the POW camp for German in the Netherlands Indies

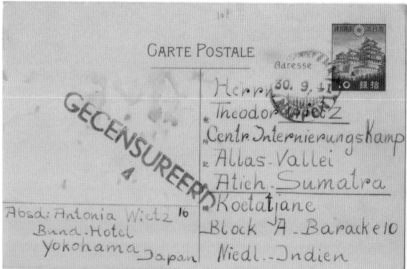

fig.22：開戦前最末期の日泰航空でのタイ宛航空書状
Latest usage of airmail to Thailand by Japan-Thai Airways before the outbreak

その他の地域

アフリカ・南米方面

　グラフィックデザイナー、小説家の妹尾 河童氏が幼少期の記憶を記述した自伝的小説「少年H」に次のような記載があります。これは、当時欧州から避難してきたユダヤ人難民について記述したものですが、1941年4月の出来事として、「その人たちは、ポーランドからシベリア鉄道に乗りウラジオストクを経て神戸へ辿りついた５３人の一団だった。」「１５日後に神戸港から、大阪商船の『まにら丸』で出港し、４０日がかりでアフリカの南端のケープタウンまで行くそうや。」との記述があります。

　この時期、有名な杉原千畝の"命のビザ"などにより、欧州を脱出したユダヤ人難民は、シベリア鉄道を経由して日本の地に至り、その後、アメリカなどの各国に行き着いています。

　図23は、まさに「少年H」に記載されたユダヤ人難民による使用例であり、ポーランドからシベリア鉄道で避難してきた彼らが、まにら丸にて南アフリカに向かう船中で差し出したものです。船内差し出しということで、「PAQUEBOT」印が押され、南アフリカで消印を押捺、アメリカに宛てられています。

　なお、本差出人である、"ABRAMOWICZ BERNARD"および、"SAMUEL"氏の素性について調査していたところ、意外なところで両氏の名前を発見することができたのです。それは、杉原千畝によるビザ発給リストに両氏の名前が並んで記載されていたのです。時期や経緯についても概ね一致しており、この葉書の両氏が、杉原千畝のビザによって脱出したユダヤ人難民であると推定され、この点においても非常に興味深く、かつ、貴重な資料と言うことができるでしょう。

Other Regions

Africa and South America

Kappa Senoo, a graphic designer and novelist, wrote an autobiographical novel "A Boy Called H." It has a description of the Jewish refugees, who escaped from Europe then, and he wrote as what happened in April 1941, "they were a group of 53 members, who came from Poland by the Trans-Siberian Railways to Vladivostok, and finally got to Kobe." And, "they said, 15 days later, they would leave Kobe by "MANILA MARU" of O.S.K. and voyage 40 days to Cape Town, at the southernmost part of Africa."

The Jewish refugees, who got away from Europe by the famous "visas of life" of Chiune Sugihara or other ways, arrived in Japan via the Trans-Siberian Railways and went to other countries such as the U.S.

fig.23 is a usage of Jewish refugees just written in "A Boy Called H." Jew who got away from Poland via the Trans-Siberian Railways sent the postalcard in MANILA MARU, on the way to Africa. Since it was sent on a ship, it received a "PAQUEBOT" mark. Then, it was canceled in South Africa and sent to the U.S.

I found the senders' names, "ABRAMOWICZ BERNARD" and "SAMUEL" in an unexpected document. It was a list of issued visas by Chiune Sugihara. The period and story almost corresponded. So, the senders were possibly Jewish refugees who escaped using the visas issued by Chiune Sugihara. It is a fascinated and valuable item on this point.

fig.23：杉原千畝のビザを受けたユダヤ人避難民発の葉書 Postalcard sent by the Jew who received the visa issued by Chiune Sugihara

この葉書は、前述の通り、1941年6月にまにら丸にて南アフリカに逓送されていますが、このあたりがアフリカ方面に向けた最末期の日本船となるでしょう。

さらにアフリカ方面について、到着便の末期使用例も図示したいと思います。図24は、南アフリカ・ケープタウンから横浜宛の使用例であり、1941年8月28日消印が押捺されています。航空路について、日泰航空経由でアフリカに宛てることは、1941年に入った頃はまだ可能でした。

南米方面について、末期の使用例をご紹介します。図25は、南アメリカ方面宛末期の使用例であり、神戸から1941年9月11日消印でブラジルに差し立てられています。裏面に到着印があり、到着は太平洋戦争開戦後の1942年3月となっています。南米方面について、ブラジル移住船が運行されており、その最終便は1941年6月22日に神戸を出港したぶえのすあいれす丸でした。さらに、開戦後最後に帰港した商船は、昭和16年10月に国の命令で銅鉱石を積取りにチリに向かった鳴門丸であり、その目的を無事に果たして翌17年1月3日横浜港に帰港しています。

This postalcard was, as described before, sent in June 1941 from MANILA MARU to South Africa, and it was one of the latest usages delivered to Africa by a Japanese ship.

Here I have the latest usage of arrival from Africa. fig.24 was from Cape Town (South Africa) to Yokohama, and it had a cancellation of August 28, 1941. The route to Africa by Japan-Thai Airways was still available at the beginning of 1941.

fig.25 was sent to South America just before the War. It was sent from Kobe on September 11, 1941, to Brazil. It has the arrival mark on the back showing the date of March 1942, after the outbreak of the Pacific War. The emigration ships were operated to Brazil, and the last ship was BUENOS AIRES MARU, which departed Kobe on June 22, 1941. The last merchant ship came back to Japan was NARUTO MARU, which left in October 1941 to Chile to carry copper ores under orders of the government. It carried outperform and returned to Yokohama the next year, on January 3, 1942.

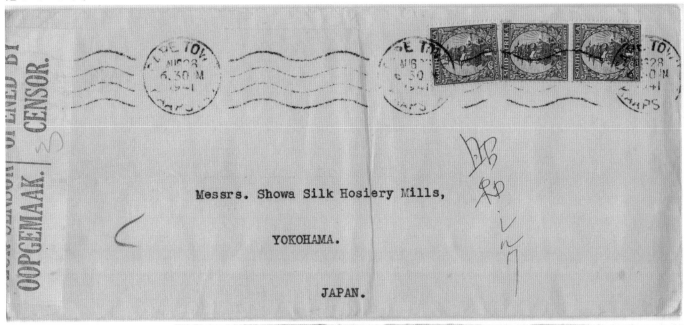

fig.24：末期のアフリカ発日本宛使用例
From Africa to Japan just before the War

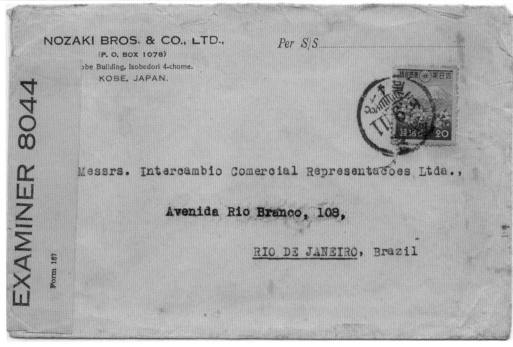

fig.25：末期の日本発南米宛使用例。到着は1942年3月。
From Japan to South America just before the War. Arrived in March 1942.

中国方面

图 26 は、1941 年 12 月 1 日に神戸からマカオに宛てられた書留郵便です。マカオは近隣国であり、この時期でもルート上の制約はあまり受けずに送達が可能だったようですが、開戦一週間前のこの時期に宛てられた外国郵便自体が珍しく、貴重な資料と言えるでしょう。表面には神戸のポルトガル大使館の印、および、現地の検閲印が押捺されています。

また、大陸方面については、平面便は戦中も送達が可能でしたが、航空郵便については著しく制限を受け、軍事および、公用での利用に限定されることとなりました。この扱いは、植民地である、台湾や朝鮮も同様であり、1941 年 12 月 12 日以降は民間人の一般利用での航空郵便送達は不可となりました。

図 27 は、1941 年 12 月以降の、公用使用に限定されていた期間の上海宛航空郵便であり、極めて貴重な使用例となります。東京から上海のスイス領事館に宛てられたものです。

China

fig.26 is a registered mail sent from Kobe on December 1, 1941, to Macau. Macau is a neighbor country, and there was not a limitation of the delivery route even at that time. But by considering the international mail itself was rare a week before the starting of the War, it is a valuable item. It has a seal of the Portuguese Embassy in Kobe and a censorship mark at arrival on the front.

The route to the Chinese Continent by ship and land was available during the War. But airmail was strictly controlled and limited for military or official use. The situation was the same in occupied territories such as Taiwan or Korea, and the airmail for private use was prohibited from December 12, 1941.

fig.27 is quite a valuable usage of airmail to Shanghai after December 1941, during the period of official-use limitation. It was from Tokyo to the Swiss Embassy in Shanghai.

fig.26：1941 年 12 月の日本発マカオ宛書状
Letter from Japan to Macau, in December 1941

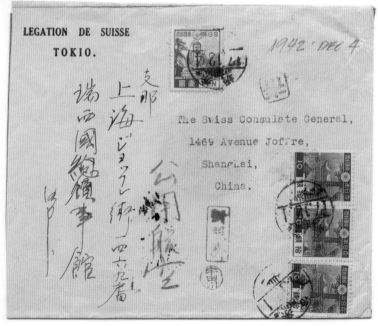

fig.27：公用のみ認められていた時期の大陸宛航空便使用例
Airmail to China in the period of official use only.

通信途絶の影響を受けた郵便など

Mails affected by the suspension of correspondence

遞送方法が途絶した郵便は当然ながら差出人に返戻されることとなります。記述してきた通り、戦争が拡大する中、細い糸を縫うように、さまざまな遞送路をもって郵便の送達が続けられてきましたが、いよいよ途絶するケースも増加し、太平洋戦争開戦をもって、多くの国との通信が完全に途絶えることとなりました。

図28は、1941年12月1日に横浜から蘭印に差し立てられた姫路城連合葉書の航空便であり、貴重な使用例です。直後の太平洋戦争開戦により、現地まで到達できず差出人に戻されたものです。日泰航空経由であり、外国の返戻印が押捺されていますので、タイまでは到達できたのかもしれません。返戻の際の、1942年2月の消印が押捺されています。

When the delivery method was suspended, of course, the mail returned to the sender. As I described before, despite the escalation of the War, the mail delivery continued by various routes, as threading the way through a very narrow gap. However, finally, the delivery suspension cases increased, and when the Pacific War broke out, the correspondence with most countries completely stopped.

fig.28 is an essential usage of airmail, a U.P.U. postalcard of Himeji Castle design, from Yokohama on December 1, 1941, to Dutch Indies. By the outbreak of the Pacific War, it couldn't reach the address and returned to the sender. Japan-Thai Airways delivered it. Since it has a foreign return mark, it might have reached Thailand. It has a cancellation at the return of February 1942.

fig.28：開戦により返戻となった、蘭印宛航空葉書
Airmail to the Netherlands Indies which returned because of the outbreak of the War

また、図29を見てみましょう。本例は、1941年11月24日消印で、米国から台湾に宛てられた郵便ですが、開戦わずか二週間前のこの時期には日本への送達は不可能であり、差出人戻しとなっています。表面に大型の返戻印がおされるとともに、図30のような付箋が添付されていました。「敵国および敵国占領地に宛てられた郵便のため返戻します」等々記載があります。

fig.29 is a mail from the U.S. dated November 24, 1941, to Taiwan. Since it was only two weeks before the starting of the War, they couldn't carry the mail to Japan and returned to the sender. It has a large return mark and a label of fig.30, which includes the description of "return because of the mail to enemy countries or their territories."

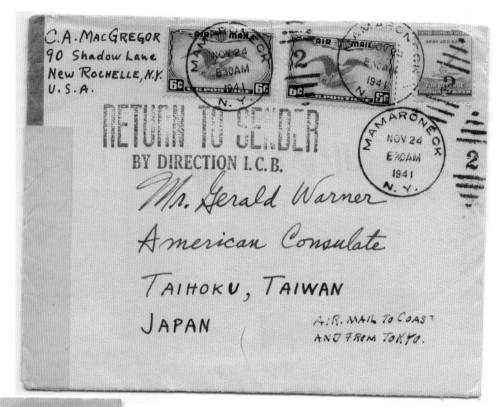

**fig.29：米国発台湾宛差し戻し便
Returned mail from the U.S. to Taiwan**

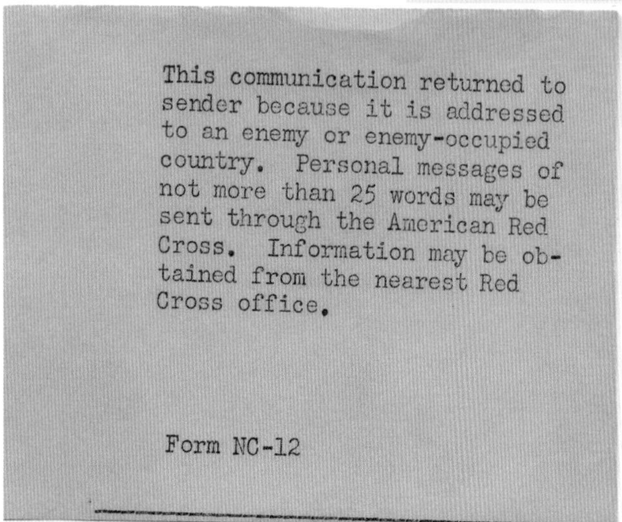

This communication returned to sender because it is addressed to an enemy or enemy-occupied country. Personal messages of not more than 25 words may be sent through the American Red Cross. Information may be obtained from the nearest Red Cross office.

Form NC-12

fig.30：同付箋 the label of fig.29

最後に非常に興味深い実例を二点ご紹介して、本稿を終えたいと思います。図31は、消印が薄く、また、書き込みがあるものの、こちらも非常に薄く分かりづらいのですが、1941年6月14日に東京に宛てられた郵便です。しかしながら、時期的に独ソ開戦の影響を受けた模様で、いずれかの地に滞留され、戦後になり日本に送達されたものです。表面に押捺された「貼付切手完全確認」の印は、切手剥ぎ取りが発生したため、到着の際に切手貼付の確認とした押されたものです。

しかしながら名宛人を確認できず、差出人に戻されたものです。1948年と薄く記載があり、差し出し後、7年あまりを経て、差出人に戻ったという興味深い例となります。

In conclusion, I show here two very interesting usages. fig.31 is, even hard to read the faint address, a mail to Tokyo, which was sent on June 14, 1941, by reading a faint cancellation and some notes. It seemed to be affected by the outbreak of the German-Soviet War. It was left undelivered somewhere and carried to Japan after the War. The postmark of "Completely prepaid by stamps" on the front was used during the postwar period, meaning the confirmation of the stamps at the arrival in case of stamps were missing.

However, the addressee of this letter couldn't be found and returned to the sender. It has faint writing of 1948, so it is an item of great interest that returned to the sender after seven years from the posting.

また、図32も同じような使用例です。本例は、図31と逆に日本から宛てられた使用例です。1941年2月にフランスに宛てられた使用例ですが、現地で名宛人が確認できず差出人戻しの扱いとなりました。

しかしながら、日本までの返戻ルートが途絶したか、ナチス侵攻の混乱の中、いずれかの地で滞留され忘れられたかなどし、終戦後にようやく返戻されたものです。

日本において、GHQの検閲がなされ、1947年5月付の付箋が貼付されています。差出人は、富山・高岡商業高校のドイツ人教師であるニコラス・ボックであり、高岡商業高校にて神戸転送の付箋を貼られ、同地へ転送されています。本例も、1941年の差し出しから6年を経て、ようやく差出人の元に帰った例であり、戦争に翻弄された郵便として、非常に貴重な資料と言えるものでしょう。

fig.32 is a similar case. This usage was the reverse route of fig.31, from Japan. It was sent to France in February 1941, but the addressee couldn't be found and returned to the sender.

However, since the return route to Japan suspended or it was left somewhere in confusion provoked by the Nazi's invasion, it returned finally after the War.

G.H.Q. censored it in Japan and put a label of May 1947. The sender was a German teacher Nicolas Boch, who had worked at Takaoka Commercial High School in Toyama. It was transferred to Kobe, attached a transfer label at the high school. It is also a return mail taking a long time of six years from the mailing in 1941. It is a vital document as a mail affected by the War.

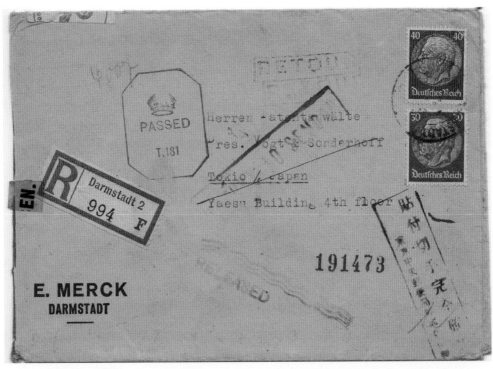

fig.31：1941年ドイツ発書状、戦後、ドイツに返戻 Letter from Germany in 1941. Returned to Germany after the War

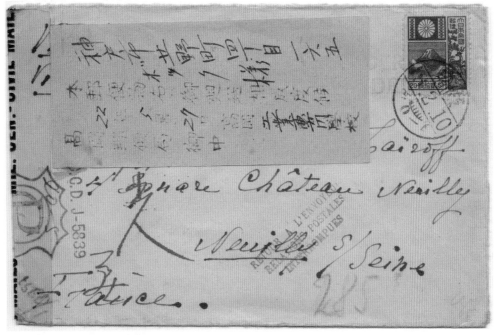

fig.32：1941年日本発フランス宛書状、戦後、日本に返戻 Letter from Japan to France in 1941, returned to Japan after the War

参考文献　References

郵趣研究（46号,50号,80号 日本郵趣協会）/ The Philately Studies vol.46,50,80, (JPS)

日本の航空郵便（成田 弘：日本郵趣協会）/ Japanese Airmail (Narita Hiroshi, JPS)

航空郵便沿革史（郵政事業史論集 第1集 山口 修：ぎょうせい）A history of Airmail (Yamaguchi Osamu, Gyosei)

少年H（妹尾 河童：講談社）SYONEN H (Seo Kappa, Kodansya)

wikipedia（https://ja.wikipedia.org）

杉原千畝記念館ホームページ Sugihara Chiune Memorial Museum（http://www.sugihara-museum.jp）

The Author： IWASAKI Tomoyuki

Collecting：Postal Histories of Japan and related areas

Awards：

   ADVERTISED POSTMARKES in JAPAN

     90pt+SP（PRAGA 2018）

     G +SP(All Japan Stamp Exhibition 2018, national exhibition)

   International Mails at reduced rate of Japan 1947-1959

     V (JAPEX 2015, national exhibition)

Membership

  Society for Promoting Philately (Japan)

  Izumi Stamp Study Group

青島局と山東鉄道沿線局の郵便史
日本租借時代：1914 年〜 1922 年
Postal History of Qingdao and the Shandong Railway Zone
Japan Leased Period: 1914 to 1922
福田真三 FUKUDA Shinzo

日独戦争（青島）

日本は大正 3（1914）年 8 月 23 日ドイツに宣戦布告し、9 月 2 日に山東半島のドイツ軍を攻略するために独立十八師団を組織し、竜口、労山湾から上陸を開始しました。その後、青島に集結したドイツ軍を包囲する基地を次々と設置し、更に、ドイツ軍の補給路を断つために山東鉄道を占拠しました。これら 16 か所（出張所を除く）の基地に野戦局を開局し、郵便物の取扱を行いました。野戦局は番号で表され、例えば、第一局は竜口→王哥庄→西哥庄→沙子口、第二局から第五局は fig.2-1.2 に示す通り、短期間に移動しました。第六局〜第十六局の開局地は更新されませんでした。

一方、海軍は第二艦隊が膠州湾を封鎖するために出動し、労山湾に停泊した船内に艦船郵便所を設置しました。10 月 31 日より青島のドイツ軍への攻撃を開始し、11 月 7 日にドイツは降伏しました。これにより、膠州湾租借地と山東鉄道沿線の権益が日本に移管されました。

Japanese-German War (Qingdao)

After Japan declared war against Germany on August 23, Taisho 3 (1914), they organized the 18th Division (Imperial Japanese Army) on September 2 to defeat the German forces in the Shandong Peninsula and started landing on Longkou and the Laoshan Bay. Then, they established base camps one after another to surround the German forces gathered in Qingdao and occupied the Shandong Railway to cut German lifeline. They established field POs other than local branches to handle the mails in these 16 base camps. Each field PO had a number, and its location changed in a short time. For example, the 1st PO changed in the following order: Longkou >Wangzhuang > West Wangzhuang > Shazikou, and from the 2nd to 5th POs as in fig.2-1.2. From the 6th to 16th POs didn't change.

The navy sent the 2nd fleet to block the Jiaozhou Bay, and they established a fleet PO on an anchored fleet at the Laoshan Bay. On October 31, they started an attack against the German forces in Qingdao, and the German forces surrendered on November 7. Japan had got the leased territories in the Jiaozhou Bay and the right of Shandong Railway zone by this war.

日独戦争（青島）の日本軍の動き Japanese Troops' Movement in the Japanese-German War (Qingdao)

fig.2-1-1

野戦局開局地 FPO

青島周辺の野戦局開局地
FPO around Qingdao

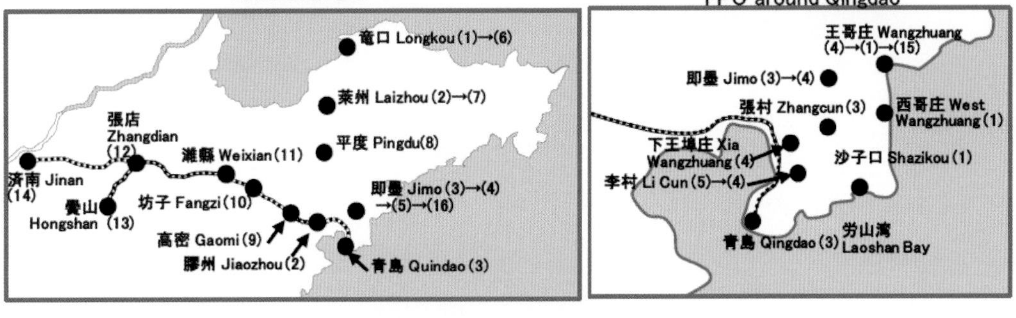

第二野戦局 2nd FPO
　莱州→膠州 Laizhou → Jiaozhou
第三野戦局 3rd FPO
　即墨→張村→青島
　Jimo → Zhangcun →Qingdao
第四野戦局 4th FPO
　王哥庄→即墨→下王哥庄→李村
　Wangzhuang → Jimo → Xia Wangzhuang → Li Cun
第五野戦局 5th FPO
　即墨→李村
　Jimo→ Li Cun

()数字は野戦局番号、矢印は野戦局番号の変遷を示す。
The number in () is a FPO number, and the arrows are the change of the number.

fig.2-1-2

当時中国の他地域の日本局では国内で発行され
ていた切手に"支那"加刷、官製葉書に"支那"字
入りが為替差益防止の目的で使われていました。
台切手は菊、大正白紙でしたが、模造防止のため
大正3年5月から旧大正毛紙切手が発行され、こ
の切手が主に租借時代に使われました。菊・大正
白紙切手と"支那"加刷のない切手の使用も確認
されています。fig.2-1.3に大正白紙、菊切手の使
用を示します。青島局で大正4年10月19日に
受付けられた芝罘宛の封書です。裏面に芝罘局の
10月23日の着印があります。

郵便料金は国内扱い封書2倍重量便6銭で、菊
1銭+1½銭+3銭+大正白紙½銭"支那"加刷の合
計6銭でした。

模造防止のため前年に旧大正毛紙が発行されま
した。大正白紙切手の発行期間が短いこともあり、
大正白紙切手の使用例は僅かです。高額の切手は
電信使用の例が多くみられます。

葉書は、国内宛に紫分銅・青分銅、国外宛に、
うす墨連合、桜連合に"支那"字入りが使用され
ました。また、封緘はがきに"軍事"加刷も使わ
れました。

At the other Japanese post offices in China then used the
stamps and postcards printed in Japan with "China" overprint
to prevent a foreign exchange gain. The base stamps used the
Chrysanthemum series and the Taisho white paper series,
however, in May Taisho 3 (1914), they changed them to the
Taisho granite old die series in order to prevent to imitate
the stamp, and this stamp was mainly used in the Japan
leased period. The usages of Chrysanthemum series, Taisho
white paper series and the stamps without "China" overprint
also have been known. fig.2-1.3 shows a usage example with
Chrysanthemum series and Taisho white paper series. It is a
letter sent to Zhifu, accepted at the Qingdao PO on October 19,
Taisho 4 (1915). It had an arrival mark of the Zhifu PO dated
October 23 on the backside.

The postage was 6 sen for double-weight domestic letter,
consisting of a 1-sen stamp with "China" overprint, a 1 ½-sen
and a 3-sen stamps of Chrysanthemum, and a ½ sen of Taisho
white paper.

The Taisho granite old die series were issued in order to prevent
imitation the previous year. usageas of the Taisho white paper
series are uncommon as the sale period was short. High face
value stamps were normally used as telegraph.

The postcards were purple or blue Fundo (shape of a weight,
symbol of wealth) design for domestic mail, and grey or cherry
blossom UPU with "China" overprint for international mail.
The "Military" overprinted postcards were also used for sealed
postcard.

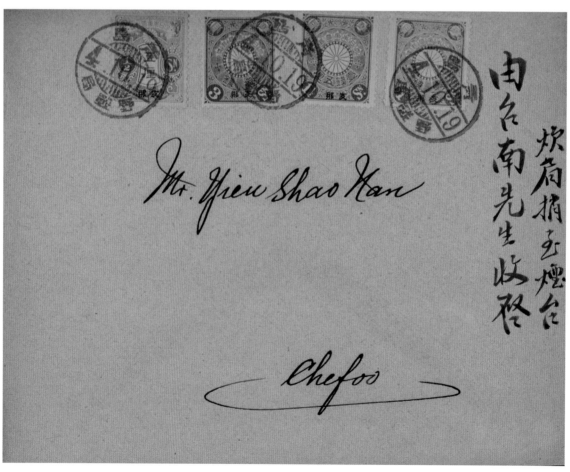

fig.2-1-3

郵便料金 / Postage

租借期間中の郵便料金は、日本国内と同じでした。また、郵便料金無料の軍事郵便制度が大正 10 年 3 月 31 日まで適用されました。そのために切手が貼付されていない葉書、封書が目立ち、面白みが少ないことが挙げられます。

During the Japan leased period, they applied the same postage of the home country Japan. And the postage-free military mail system was applied until March 31, Taisho 10 (1921). You will find frequently the postcards and letters without stamps during that period. It's less fun.

国 内 Internal Mail				
軍事郵便 Military Post	第一種 1st Class	第二種 2nd Class	第四種 4th Class	書留 Registration
	有封書状 Sealed Letter	葉書 Postcard	印刷物など Printed Matter, etc.	
1914.8.27 ～ 1921.3.31 無料 free	1899.4.1 ～ 4 匁毎 3 銭 3sen /4monme	1899.4.1 ～ 1 ½ 銭 1 ½ sen	1889.10.1 ～ 30 匁 2 銭 2sen / 30monme	1900.10.1 ～ 7 銭 7sen

国 外 International Mail				
第一種 1st Class		第二種 2nd Class	第四種 4th Class	書留 Registration
有封書状 Letter		葉書 Postcard	印刷物など Printed Matter, etc.	
1897.10.1 ～ 15g 毎 10 銭 10sen/15g		1897.10.1 ～	1897.10.1 ～	1897.10.1 ～
1907.10.1 ～ 10 銭、6 銭増 10 Sen, 6 Sen additional rate		4 銭 4sen	50g 毎 2 銭 2sen/50g	10 銭 10sen
1922.1.1				
20 銭、10 銭増 20 Sen, 10sen additinal rate		8 銭 8sen	4 銭 4sen	20 銭 20sen

fig.2-1-4

櫛型郵便印各部の呼称
Section of Comb Cds

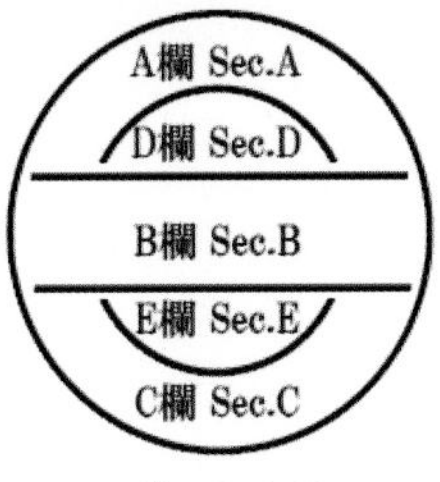

fig.2-1-5

郵便印 / Postmark

日本租借時代に使用された郵便印は、記念特印、機械印を除き全て櫛型印で、直径は全て 24.5mm、A 欄：野戦番号、地名、B 欄：年月日、E 欄：櫛型でした。fig.2-1.5 に櫛型印の各部（A ～ E 欄）の呼称を示します。

In the Japan leased period, all the postmarks, but commemorative postmarks or machine cancellations, were comb style cancellations. All of them were 24.5 mm in diameter and they had a field PO number or location name (Sec. A), date (Sec. B) and comb pattern (Sec. E). fig.2-1.5 shows each section name (Sec. A to E).

野戦局に普通局が併置され、D 欄の表示が " 野戦 " → " 櫛型 " に変更されました。また、濰縣・済南を除く租借外の野戦郵便局は大正 9 年 1 月 29 日に軍事郵便局に改称され、変更時期に併せ順次、C 欄の表示が " 野戦 " → " 軍事 " と変更されました。

A field PO was including an ordinary PO and the ordinary post office used a postmark with "comb pattern" in the Sec. D, changed from "field operations." The field POs out of the leased territories but the Weixian PO and the Jinan PO, changed the name to "military post office" on January 29, Taisho 9 (1920), and at the same time the Sec. C of a postmark also changed from "Field operations" to "Military."

逓送路 / Delivery Route

青島～日本：日独戦争中の郵便は、各野戦局から沙子口野戦局に送られました。そして、労山湾に停泊していた船舶に積込まれ、佐世保軍港入港後、佐世保駅経由で各地に逓送されました。戦争終結後の大正 3 年 12 月末から民間汽船会社（日本郵船、大阪商船、原田商行：原田汽船の 3 社）による定期航路が開設され、門司、下関細江局が交換局となり、各地に逓送されました。列車による輸送は通信省（郵便結束図に明記）により指定された列車で運ばれました。

Qingdao to Japan: The mails were sent from each field PO to the Shazikou Field PO during the Japanese-German War. Then, the ship in the Laoshan Bay took on them and delivered to the Sasebo naval port, and the mails were delivered transiting the Sasebo station to the addressee. After the War, from the end of December in Taisho 3 (1914), the private steamship companies (three companies of Nippon Yusen, Osaka Mercantile Steamship and Harada Shoko/Harada Steamship) established the regular service, and the Moji PO and the Shimonoseki-Hosoe PO became exchange POs to deliver other places. The Ministry of Communications designated the railway to deliver mails (specified in the mail network map).

青島～各国：シベリア鉄道は第一世界大戦とロシアの政権交代で逓送が不透明となりました。この影響により欧州宛の郵便は日本からアメリカを経由する逓送が増加しました。第二次世界大戦まで日本の汽船会社が活躍する時代と言えます。

Qingdao to Other Countries: The delivery by the Trans-Siberian Railway became unclear because of the WWI and the change of government in Russia. It affected the mail delivery from Japan to Europe, and the mails via U.S. were increased. Japanese steamship companies had played an important role until the WWII.

1. 日独戦争（青島）

初期番号入り野戦局

　戦闘中に各地に野戦局が開局されました。独立十八師団による野戦局を初期番号入りとし、戦闘終結後の大正3年12月16日から青島守備軍逓信部に引継がれ16局から12局に整理された野戦局を後期番号入りとしました。

　fig.2-1.6は第四野戦局（初期番号入り）の野戦印の使用例です。この時第四野戦局は即墨に設置され、そこから沙子口に運ばれ労山湾に停泊していた三河丸で佐世保港に運ばれました。佐世保から宛先の長興に時津局を経由して逓送されました。

1. Japanese-German War (Qingdao)

Field PO with Early Number

The field POs were established in many places during the battle. Here I call the field POs established by the 18th Division "early number," and name "late number" to those taken over to the communication division of Qingdao Defensive Army and disposed from 16 to 12 offices after the battle end on December 16, Taisho 3 (1914).

fig.2-1.6 is a usage example of the postmark of the 4th Field PO (early number). The 4th Field PO was established in Jimo then. The letter was delivered from there to Shazikou, and then by Mikawa Maru anchored in the Laoshan Bay to the Sasebo harbour. Finally, it was delivered from Sasebo to the addressee Nagayo passed by the Togitsu PO.

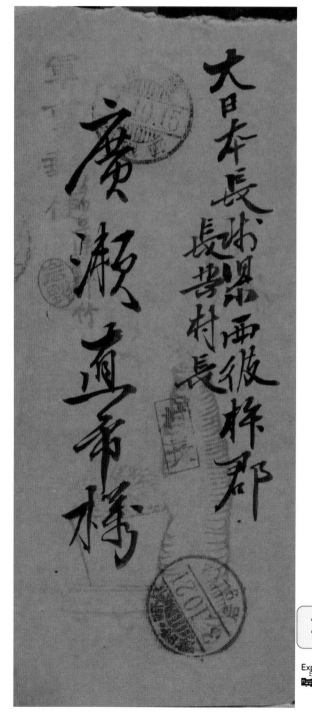

fig.2-1-6

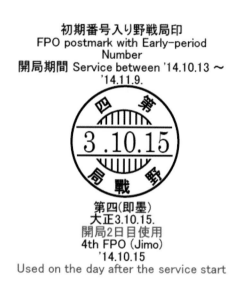

初期番号入り野戦局印
FPO postmark with Early-period Number
開局期間 Service between '14.10.13 ～ '14.11.9.

第四（即墨）
大正3.10.15.
開局2日目使用
4th FPO (Jimo)
'14.10.15
Used on the day after the service start

fig.2-1-7

逓送路 Route

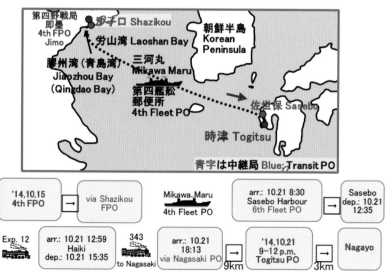

fig.2-1-8

青島での戦闘中の郵便輸送のため、労山湾沖に停泊した船舶が佐世保との間を運航しました。運行された船舶に第三・四艦船郵便所が設置され、佐世保本部に第六艦船郵便所が設置されました。艦船郵便所の開始時期は下記の通りです。

・第三艦船郵便所：大正 3 年 9 月 9 日、三池丸
・第四艦船郵便所：大正 3 年 9 月 6 日、三河丸
・第六艦船郵便所：大正 3 年 9 月 5 日、佐世保本部
（資料：「海軍日誌　大正 3、4 年役」佐世保港湾部）

fig.2-1.9 に第四艦船郵便所（三河丸）の使用例を示しました。三河丸は 10 月 21 日佐世保港を出港し、労山湾に 10 月 22 日から 30 日迄停泊し、10 月 31 日に佐世保港に帰港しました。

この私製葉書は労山湾碇泊中の 10 月 28 日に受付けられました。佐世保港に帰港した後の宛先への逓送を fig.2-1.11 に示しました。

During the battle, the ships anchored off the Laoshan Bay had mail delivery service between Qingdao and Sasebo. On the ships, the 3rd and 4th Fleet POs were established and the Sasebo head office had the 6th Fleet PO. The service start date of each fleet post offices were the follows:

·3rd Fleet PO: Sep. 9, Taisho 3 (1914), Miike Maru
·4th Fleet PO: Sep. 6, Taisho 3 (1914), Mikawa Maru
·6th Fleet PO: Sep. 5, Taisho 3 (1914), Sasebo head office
(Reference: "The Navy Diary, Taisho 3 and 4" Sasebo Port and Harbour Division)

fig.2-1.9 shows a usage example of the 4th Fleet PO (Mikawara Maru). Mikawa Maru left the Sasebo Harbour on October 21, anchored in the Laoshan Bay from October 22 to 30, and went back to the Sasebo Harbour on October 31.

This private postcard was accepted on October 28, during anchoring in the Laoshan Bay. fig.2-1.11 shows the delivery route from the Sasebo Harbour to the addressee.

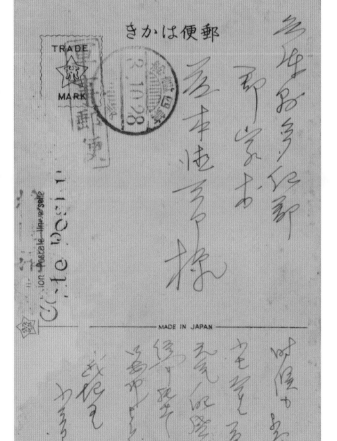

fig.2-1-9

艦船郵便所印
Fleet PO Mark
受付期間 accepted between '14.9.6.～
'14.12.24.

第四艦船(三河丸)／大正3.10.28
C欄"郵便所"
4th Fleet (Mikawa Maru)/ '14.10.28
Sec.C "HPO"

fig.2-1-10

逓送路 Route

青字は中継局 Blue: Transit PO

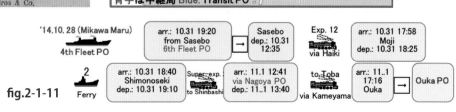

'14.10.28 (Mikawa Maru)
4th Fleet PO

fig.2-1-11   2 Ferry

俘虜郵便

POW Camp Post

山東半島の日独戦争では約 4,500 名のドイツ人俘虜が国内 16 箇所（fig.2-1.12 参照：久留米に寺院・バラック・高良内を含む）の俘虜収容所に送られました。これらの郵便物は、俘虜郵便 "SERVICE DES PRISONNIERS DE GUERRE" として無料扱いでした。

日独戦の俘虜に関しては有名な実話が残されています。坂東（徳島県鳴門市）収容所には約 1,000 名を 1917 年～ 1920 年まで収容しました。この収容所で日本では初めてベートーヴェンの交響曲第九番が演奏されました。

fig.2-1.12 に青島受付、姫路俘虜収容所宛の官製葉書を紹介します。大正 3（1914）年 12 月 17 日に第一野戦局（青島）で受付けられた姫路俘虜収容所宛の官製葉書です。

姫路俘虜収容所は大正 3 年 11 月 11 日～大正 4 年 9 月 20 日の間、俘虜の受け入れを行いました。

この葉書は比較的早い時期の俘虜郵便です。郵便料金は無料ですが、うす墨連合葉書 4 銭 "支那" 字入りの国外用の葉書が使用されました。

During the Japanese-German War in the Shandong Peninsula, about 4,500 German POWs were sent to 16 POW camps in Japan (See fig.2-1.12: Kurume camp including a temple, barracks and Kourauchi). Their mails were handled for postage free as POW post, "SERVICE DES PRISONNIERS DE GUERRE." There is a famous story of the POWs in the Japanese-German War. The Bando POW Camp (Naruto city, Tokushima Prefecture) interned 1,000 POWs from 1917 to 1920. In this camp, the Symphony No.9 of Beethoven was played for the first time in Japan.

fig.2-1.12 was a postcard to the Himeji POW Camp, accepted at Qingdao.

The 1st Field PO (Qingdao) accepted it on December 17, Taisho 3 (1914) and sent to the Himeji POW Camp.

The Himeji POW Camp interned POWs between November 11, Taisho 3 (1914) and September 20, Taisho 4 (1915).

This postcard is a POW mail of the relative early period. The postage was free, but the grey UPU 4 sen international postcard with "China" overprint was used.

fig.2-1-13

fig.2-1-12

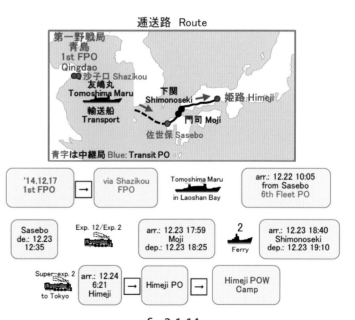

fig.2-1-14

　大正4年4月1日に郵便印の番号入り野戦局表示が地名入り野戦局表示に変更されました。fig.2-1.15に4月7日に青島局で受付けられたデュッセルドルフ宛の使用例です。郵便印は番号入り野戦局印とA欄の活字が変更されているだけで、他の変更はありません。青島の日本軍の検閲を受け、検閲済の赤色の角印が押印されています。

　郵便料金は国外宛4銭により、うす墨連合葉書4銭 " 支那 " 字入りが使われていました。この葉書は、日本を経由して太平洋航路でシアトルに送られました。門司、横浜局の中継印が押印されています。当時の太平洋航路は、日本郵船、大阪商船がシアトルから、東洋汽船がサンフランシスコからアメリア鉄道と連絡をしていました。シアトルからのアメリカ鉄道は Northern Pacific Lines（北回り）、サンフランシスコからは Southern Pacific Lines（南回り）でシカゴに送られました。どちらも3日間を要しました。シカゴからニューヨークは New York Central Lines、Pennsylvania Lines のどちらかで送られました。

　ニューヨークから大西洋航路のドイツ宛は第一次世界大戦前は Lloyd 汽船、H-Aline といったドイツ系の汽船会社が主力でしたが、大戦後はオランダ、ベルギーの汽船会社が主力となり、ロッテルダム経由でドイツ宛となりました。シベリア鉄道の利用が避けられ、日本、アメリカ経由のヨーロッパ宛が増加した時期でした。

On April 1, Taisho 4 (1915), a field postmark with number changed to with location name. fig.2-1.15 is a usage example to Düsseldorf, accepted at the Qingdao PO on April 7. The difference between field postmark with number and this one was only the description of Sec. A. It was censored at the Japanese Army in Qingdao, and it had a red rectangle censored mark.

The postage was 4 sen for international mail, and a grey 4-sen UPU postcard with "China" overprint was used. This postcard was delivered via Japan by the Pacific route to Seattle. It had transit marks of the Moji PO and the Yokohama PO. The service over the Pacific then were connected to the U.S. railway from Seattle by Nippon Yusen and Osaka Mercantile Steamship, and from San Francisco by Toyo Kisen. The U.S railway delivered the mails from Seattle, by the Northern Pacific Lines (the northern route) or the Southern Pacific Lines (the southern route) from San Francisco, to Chicago. It took three days by either line. From Chicago to New York, the New York Central Lines or the Pennsylvania Lines was used.

The mails from New York to Germany by the Atlantic route was mainly operated by German origin companies such as Lloyd or H-Aline before the WWI, and in the postwar period Dutch or Belgian steamship companies mainly delivered mails via Rotterdam to Germany. It was a period when they avoided from using the Trans-Siberian Railway and the mails via Japan and U.S. to Europe increased.

fig.2-1-15

時刻表示なし野戦局印
FPO postmark without time
使用期間 used between
'15.4.1 ～ '18.8.8

.4.4.7

青 島 Qingdao
大正4（'15）.4.7

fig.2-1-16

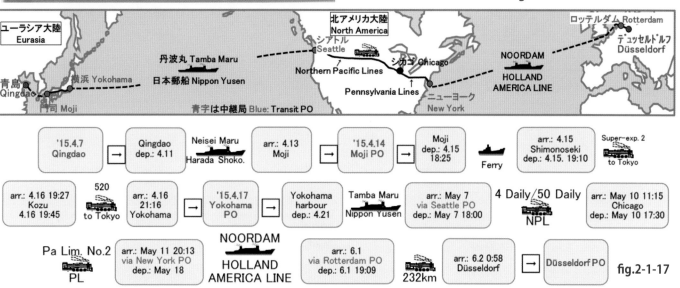

fig.2-1-17

次に中国のフランス領宛の使用例を紹介します。この私製葉書はフランス占領地の広州湾 FORT-BAYARD（現在の湛江）宛で、中国郵政との連携をせず日本の逓送ルートで門司・上海を経由して香港に送られ、香港からフランス船で FORT-BAYARD に送られました。郵便料金が国外宛葉書4銭により菊2銭 "支那" 赤加刷が2枚貼られました。

青島から中国郵政による上海経由香港宛も可能であったとかんがえますが 日本の逓送で香港まで送られ、フランス船により FORT-BAYARD に送られています。"Fort" とは砦との意味です。葉書に "門司廻し" と青字で書かれています。

The next is a usage example to French settlement in China. This private postcard was sent to the French settlement FORT-BAYARD (known as Zhanjiang today) in the Guangzhou Bay, delivered without cooperation with Chinese postal service but using the Japanese route, via Moji and Shanghai to Hong Kong, and from Hong Kong by a French ship to FORT-BAYARD. The postage was 4 sen for international postcard and it bears two 2-sen Chrysanthemum stamp with "China" overprint.

It would have been possible to deliver by the Chinese postal service from Qingdao via Shanghai to Hong Kong, but in this case, it was delivered by Japanese service to Hong Kong, and by a French ship to FORT-BAYARD. The postcard had "transfer to Moji" description in blue ink.

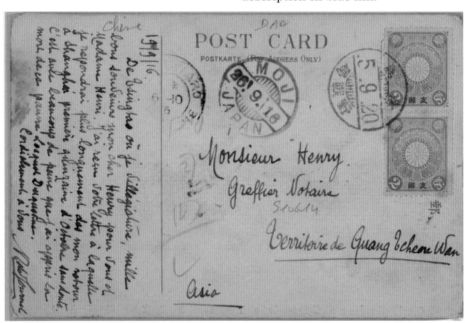

fig.2-1-18

時刻表示なし野戦局印
FPO postmark without time
使用期間 used between '15.4.1〜 '18.8.8

青島 (Qingdao)/大正5 ('16).9.20.

fig.2-1-19

逓送路 Route

fig.2-1-20

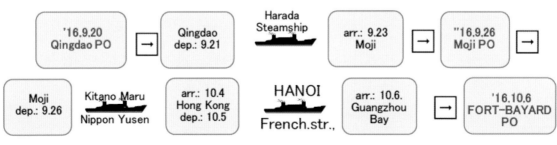

fig.2-1-21

クリックで選択

Cliquer et choisir

Haz clic y elige

點擊選項

Click and pick

Нажмите и выберите

Clicca e scegli

クリックで選択

Vaccari
chaponnière & firmenich　Il Ponte　Eisenhammer　Bendon
Colonial Stamp Co　Kelleher Rüstkammer　Siegel Dorotheum
stampBAY Badisches Auktionshaus　Schwarzenbach　Hadersbeck Nomisma
Victoria Stamp Co.　　　　　　　　　　Leiningen
Schatztruhe　John Bull Höhn
Drei Löwen　Philatino MAR
Gert Müller　Roumet HP Balkanphila
Soler y Llach　stampedia Bielefelder
Pilatusmail.ch　Dr. Derichs Köln Romano
Kupersmit Sommer Geigle
Weiser Iberphil　Hettinger Juranek Dessau
Athens J&K　Heickmann Mayfair
Bolaffi Eppli　Cavendish Südphila
Morrison　AB Philea Helvetia
SFP Group DBA　La Postale Tosunidis
UPA　Teutoburger Tel Aviv
Nordphila Lenz　Sammarinese Kempf
Spink Van Dieten Toselli　Honegger AG Rainey
Casati　Doreen Royan Hodam
Fickert Corinphila Stilus　Stanley Gibbons Pfankuch
Michael Rogers scriposale　Eastern
Llach cedarstamps　HWPH Postiljonen
Asbit Yves Siebers　Ferrario
Callies　Numphil Corinphila NL
Dreyfus Philangles Mirko Franke　Wohlfeil Rölli
Bach　Filat Baumeister Götz　Fischer Geier
D&T Intl.　Lippolds Zofingen
Chris Rainey　Boule AAK Phila　Bloxham Bühler
Auction Gallery Kiefer　Heuberger Corbitts
Cherrystone　Philadria　abacus Rijnmond Christ
Deider　Merkur Schlegel　Viennafil Cérès WAP
Cortrie Rapp Forster Tiergarten　Badische Bfm. pastbuy
MH Marken　Lugdunum　Harmers SA Köhler
Burda Harlos Mohrmann Interasia Behr Argyll Etkin
Auction Phila Gärtner Kirstein-Larisch Felzmann
Rauss & Fuchs HBA
Veuskens
Schulz David Feldman
F. Feldman classicphil
Arbeiter PhilAuction
Stade CompuStamp
Dresdner Aix-Phila
Raritan Grosvenor
Laser Invest Plumridge
Schuyler Rumsey
Complete Stamp Embassy
Böhmenphila Schantl
Hörrle
Roumet Auction Galleries HH
Rauch Pilatusmail Auktion
Kelleher & Rogers
Rauhut & Kruschel
Uwe von Poblocki
Dr. Fischer
Honegger M.
Merkurphila
Phila China

www.philasearch.com

　租借時代に各野戦局で国内と同様に記念特印が使われました。fig.2-1.22 に青島局で使われた記念特印を示しました。この封書は、大礼奉祝会場内に青島野戦郵便局臨時出張所が設置され、スウェーデン宛の封書に特印が押印されました。受付は 4 日間でした。fig.2-1.23 が封書裏面のコピーで、門司の中継印とストックホルムの着印が押印されています。郵便料金は国外宛封書 10 銭により大正毛紙 10 銭 " 支那 " 加刷が使われました。

In the occupation period, field POs used commemorative postmarks as in the home country Japan. fig.2-1.22 shows a commemorative postmark used at the Qingdao PO. This letter was sent to Sweden with a commemorative postmark which was put at the provisional branch of the Qingdao Field PO established in the celebration hall for Taisho Emperor coronation. The accepting period was 4 days. fig.2-1.23 was the backside of the letter and it had a transit mark of the Moji PO and an arrival mark of the Stockholm PO. The postage was 10 sen for international letter, bearing a 10-sen stamp of the Taisho granite with "China" overprint.

門 司 Moji
'15.11.17

↓

ストックホルム Stockholm
'15.12.30 11-12

fig.2-1-23

大正大礼記念特印
Commemorative Special
Postmark for Taisho Emperor
Coronation Ceremony
受付期間 accepted between
'15.11.10.〜11.14

青　島／大正4.11.12.
Qingdao/ '15.11.12

fig.2-1-22

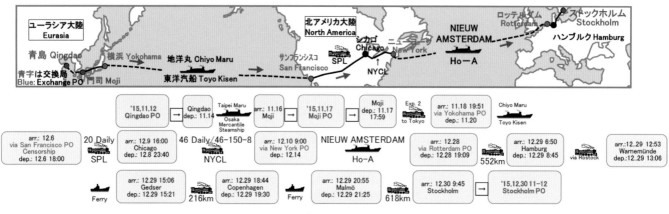

fig.2-1-24

この逓送路は、fig.2-1.24 にある通りバルト海を渡り、コペンハーゲンまで送られました。サンフランシスコで検閲を受けましたので、サンフランシスコ発は入港当日としましたが、ニューヨークで3日程度の余裕がありますので、実際には翌日、翌々日かもしれません。但し、着印から Ho－A 汽船の NIEUW AMSTERDAM で運ばれました。当時のオランダの定期航路は1回/週でした。大西洋航路の汽船の運行は New York Times のメール一覧で確認しています。

もう一つ青島局で使われた記念特印の使用例を紹介します。大正5年11月3日裕仁立太子の記念切手が販売され、特印が配置されました。fig.2-1.25 は記念印が使われたパリ宛の私製葉書です。希望者には 11月5日までの3日間特印の押印が受付られました。初日の日付です。郵便料金は国外宛葉書4銭で、菊4銭"支那"加刷切手が使われました。

この葉書は、青島から大連定期船で上海に運ばれ MM 郵船の汽船でマルセイユに運ばれました。マルセイユからはパリ着印から入港翌日の朝一の特別列車で運ばれました。

The delivery route (fig.2-1.24) in Europe was across the Baltic Sea to Copenhagen. The letter was a subject to censorship at San Francisco, so even though I described the arrival date of San Francisco as departure date, because of about three spare days in New York, possibly the actual departure date of San Francisco would be the next or the day after next. It was clear by an arrival mark that the Ho-A's NIEUW AMSTERDAM delivered the letter. The Dutch companies then had a weekly regular service. I've found the Atlantic route service in the mail list of the New York Times.

I show another commemorative postmark example of the Qingdao PO. The commemorative stamps for Crown Prince Hirohito was issued on November 3, Taisho 5 (1916), and also the commemorative postmark was prepared. fig.2-1.25 was a private postcard to Paris with a commemorative postmark. The special postmark was put if wanted for three days until November 5. This postcard had a postmark of the first day. The postage was 4 sen for international postcard and it bears a 4-sen Chrysanthemum with "China" overprint.

It was delivered from Qingdao by a regular Dalian liner to Shanghai, and then, by a MM's steamship to Marseille. The arrival mark at the Paris PO shows that it was delivered from Marseille by the first special train of the next day of arrival.

fig.2-1-25

裕仁立太子記念特印
Commemorative Special Postmark for Crown Prince Hirohito
受付期間 accepted between '16.11.3〜11.5

青島 (Qingdao)/
大正5 ('16).11.3

裏面着印コピー
Arrival Mark on the Back

PARIS-PLAGE (市内局 Local PO)
'16.12.13 7:00

fig.2-1-26

欧州逓送鉄路
Route in Europe

アジア・欧州航路(MM郵船:マルセイユ経由)
Asia-Europe Route (MM: via Marseille)

ユーラシア大陸 Eurasia

拡大 Enlarged
パリ Paris
リヨン Lyon
マルセイユ Marseille
至ポートサイド to Port Said
青字は交換局 Blue: Exchange PO

ポートサイド Port Said
スエズ運河 Suez Canal
スエズ Suez
アフリカ大陸 Africa
ジブチ Djibouti
青字は中継地 Blue: Transit Point

ボンベイ Bombay
コロンボ Colombo
MEGELLANG

青島 Qingdao
上海 Shanghai
香港 Hong Kong
サイゴン Saigon
シンガポール Singapore

fig.2-1-27

'16.11.3 Qingdao → Qingdao dep.: 11.4 — Sakaki Maru Mantetsu Liner — arr.: 11.6 via Shanghai dep.: 11.12 — MAGELLANG MM

arr.: 12.11 Marseille dep.: 12.12 10:35 — via Lyon — arr.: 12.13 6:20 via Paris PO — '16.12.13 7* Plages PO (Paris Local PO)

青島本局（時刻表示あり野戦局印）

郵便物の増加に伴い大正 7 年 8 月 11 日より C 欄時刻表示の郵便印に変更されました。

ここで、青島局で使われた郵便印を整理します。

① 時刻表示なし野戦局印
② 時刻表示あり野戦局印
③ 普通局印

このような順序になります。fig.2-1.28 に①と②郵便印を fig.2-1.29 に各カバーを示しました。

時刻表示なし野戦局印
FPO postmark without time
使用期間 used between
'15.4.1〜 '18.8.8

青 島
大正7.8.5. 最後期使用
C欄"野戦局"
Qingdao
'18.8.5 Last Usage
Sec. C "FPO"

時刻表示あり野戦局印
FPO postmark with time display
使用期間 used between
'18.8.11 〜 '18.11.19

青 島
大正7.8.12. 変更翌日使用
C欄"前9-12"、D欄"野戦"
Qingdao
'18.8.12,
Usage on the Day after Change
Sec. C "9-12 a.m.", Sec. D "FIELD
OPERATIONS"

fig.2-1-28

①の確認されている最後期使用は 8 月 8 日②の最初期使用は 8 月 11 日です。この間に変更されたことになります。

②の C 欄時刻入りの印顆は日本本国で製作され青島に送られました。8 月 10（土）日青島入港の薩摩丸で運ばれたと考えています。①から②の印顆の組み替えは 8 月 11 日（日）に行われ、翌 8 月 12 日（月）から使用が開始され、所有している封書が変更初日の使用と考えていました。ところが、8 月 11 日の使用例の連絡があり、fig.2-1.29 右は変更翌日になってしまいました。残念でした。

山東鉄道沿線局の②の確認されている最初期の年月日を fig.2-1.30 に示しました。膠州は翌年の確認ですが、概ね 10 月までの使用が確認されていますので、8 月中に②に変更されたと考えています。

fig.2-1.29 左は①の最後期 3 日前の使用です。郵便印を fig.2-1.31 に拡大しました。よく見ると "年" の活字の前にピリオド "・" があります。"年月日" の活字は 3 分割されています。"年" の活字の位置に "月" 又は "日" の活字が利用されました。このような例はよく見かけます。

Qingdao CPO (Field Postmark with Time Description)

When the mails increased, the Sec. C design changed to the time description on August 11, Taisho 7.

Here I organize the postmarks used at Qingdao PO:

(1) field postmark without time
(2) field postmark with time
(3) ordinary postmark

I put them in a chronological order. fig.2-1.28 shows the postmarks of (1) and (2), and fig.2-1.29 shows their covers.

The latest usage of (1) was dated August 8, and the earliest use of (2) was dated August 11. These two types must have been switched between these days.

The (2) postmark with time was manufactured in the home country Japan and sent to Qingdao. It was thought to have been delivered by Satsuma Maru which arrived at Qingdao on August 10 (Sat.). The switch from (1) to (2) was done on August 11 (Sun.). So, I had believed that the next day, on August 12 (Mon.) they started to use the new postmark and the letter I have was the first switched day use. However, I received the notice that there is a usage example of August 11. The letter on the right of fig.2-1.29 became the next of switched day use. It's unfortunate.

fig.2-1.30 shows the earliest use date of (2) postmark in the Shandong Railway zone post offices. Most of the recorded earliest dates of the post offices except Jiaozhou are by October 1918, the postmarks must have been switched to (2) by August.

The left postcard in fig.2-1.29 is a (1) postmark usage of three days before the latest day. You can see the enlarged postmark in fig.2-1.31. If you see it carefully, you can find a period "." before the "Year" number. "Year-Month-Day" parts were separated each other. In this case, "Month" or "Day" letter was used for "Year" letter part. It frequently happened.

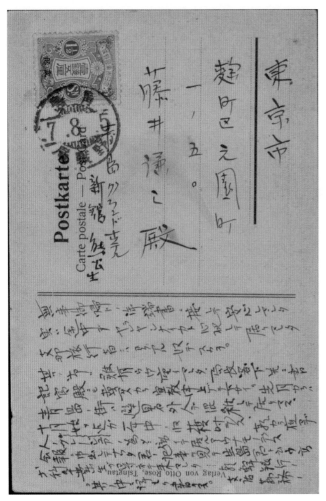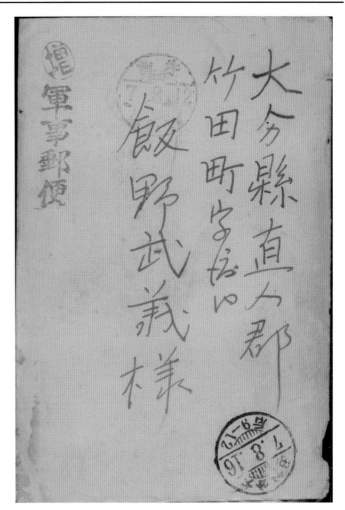

fig.2-1-29

D欄"野戦"の各局の最初期使用年月日
The Earliest Usage Date of Sec. D "FIELD OPERATIONS"

fig.2-1-30

郵便印コピー100%
Postmark

"年"の活字に
".月"又は".日"の活字使用
The ".Month" or ".Day" printing
type was replaced the "Year."

fig.2-1-31

大正 7 年 10 月 10 日に中国と「山東日支通信連絡細項取極め」が調印され、青島・濰縣・済南局が中国との交換局になり、D 欄 "櫛型" の普通局郵便印③が使われました。fig.2-1.32 に②の最後期使用と③の最初期使用を示しました。fig.2-1.33 に②、③の郵便印を示しました。

青島守備軍公報第 557 号により大正 7 年 11月 21 日より青島、青島大鮑島、青島大港、台東鎮、四方、李村、濰縣、済南の 8 局に普通郵便局が併設され、青島郵便局埠頭出張所が設置された。

そこで、11 月 21 日に③の郵便印の使用開始と考えられていました。ところがその前に使用が開始されています。

On October 10, Taisho 7 (1918), China and Japan signed the "Agreement in Detail of Correspondence between China and Japan in the Shandong Province," and the post offices of Qingdao, Weixian and Jinan became the exchange POs with China, and ordinary postmark (3), with "comb pattern" in Sec. D was used. fig.2-1.31 shows the last use of (2) and the earliest use of (3). fig.2-1.33 shows the postmarks of (2) and (3).

The Qingdao Defensive Army Report No. 557 said that on November 21, Taisho 7 (1918), an ordinary post office was attached to each eight field POs such as Qingdao, Qingdao-Dabaodao, Qingdao-Dagang, Taitungtschen, Syfang, Li Cun, Weixian and Jinan, and the Quay branch of Qingdao PO was established.

So, it had been thought November 21 was the first day of the (3) postmark use. However, the starting date was before that day.

fig.2-1-32

fig.2-1-36

C欄 "時刻"、D欄 "野戦"
Sec. C "time," Sec. D "FIELD OPERATIONS"
使用期間 used between '18.8.11 ～ '18.11.19

青島
大正7.11.19. 最後期使用
C欄 "后0-3"
Qingdao
'18.11.19 Last Usage
Sec. C "0-3 p.m."

櫛型印 C欄 "時刻"、D欄 "櫛型"
Comb Cds, Sec. C "time," Sec. D "comb pattern"
使用期間 used between '18.10.22 ～ '22.12.10

青島
大正7.10.22. 最初期使用
C欄 "前9-12"
Qingdao
'18.10.22 Earliest Usage
Sec. C "9-12 a.m."

fig.2-1-33

上葉書の郵便印の特徴
Characteristics of the above postcard
枠線との間の隙間が大きい
Large blank between the margin and letters
櫛型と横線に隙間がない
Without gap between comb pattern and horizontal line

A欄とD欄を現地にて調整（下方向にずれ）
Arranged Sec. A and D (Shifted downward)

fig.2-1-35

大正 7 年 10 月 10 日に青島で「山東日支通信連絡細項取極め」が調印されました。概要は、「日本郵政主管庁は青島に於ける支那郵便局一箇所を又支那郵政主管庁は濰縣及済南に於ける日本郵便局各一箇所を設置し従来の慣例に依り交換局として承認する。日本郵政主管庁及支那郵政主管庁は両郵政主管庁間に青島、濰縣及済南に発著し又はこれを経由する各種郵便物（普通、書留、国際及継越閉嚢又は開嚢）の定時交換を行う。」となっています。

これに基づき青島守備軍の独断専行で③の郵便印が現地で調製されたと考えています。

ここで、櫛型印の印顆の構造について説明します。fig.2-1.34 を見てください。郵便印の各部の呼称は、A 欄～ E 欄ですが、構造は、A・D 欄（上部）と C・E 欄（下部）はそれぞれ一体構造となっています。

従って、②から③の変更は、上部を替える必要があります。そこで、現地では、大正 7 年 8 月頃に①から②の変更の際に外した上部を再利用したと考えました。これであれば現地で調製が可能となります。

fig.2-1.32 の右の③の最初期使用の郵便印を fig.2-1.35 に模式化しました。上部が若干下方向にずれています。本来横 2 線と櫛（D 欄・E 欄）の間に隙間がありますが、ひっついています。これは上部が印顆に収まらず、下部を削り収めたと考えられます。

その後、11 月 21 日の普通局への移設に併せ本国から印顆が届き、組み換えが行われました。fig.2-1.36 に " 青島 " の活字が下方向にずれた印顆の使用例を示しました。新たな印顆が届いたのに現地で調製された印顆が継続して使用されています。この時期には印顆が不足していたようで興味深い印影のものが幾つか見られます。

On October 10, Taisho 7 (1918), the "Agreement in Detail of Correspondence between China and Japan in the Shandong Province" was concluded in Qingdao. Its outline was "the Japanese postal authority would establish a Chinese post office in Qingdao and the Chinese postal authority, a Japanese post office each in Weixian and Jinan. These post offices would be approved as an exchange PO according to the usual custom. The Japanese and Chinese postal authorities would regularly exchange each other the mails (ordinary, registered and sealed/opened international mails or mails crossed a border) which were sent from/to Qingdao, Weixian and Jinan or passed by these places."

According to this agreement, Qingdao Defensive Army might have prepared the (3) postmark arbitrarily on their own authority.

Now I describe the composition of a comb style handstamp. You can see in fig.2-1.34 to find the Sec. A and D (upper part) and the Sec. C and E (lower part) were as one each, even though the Sec. A to E seem to be separated parts.

Therefore, for the switch from (2) to (3), they had to change the upper part. And for there, they possibly reused the upper part removed around August in Taisho 7 (1918) when they changed from (1) to (2) postmark. They could prepare a new handstamp with this method.

fig.2-1.35 is a schematic depiction of the earliest use (3) postmark in the right of fig.2-1.32. The upper part was slightly shifted downward. It had originally a gap between each of two horizontal lines and comb pattern sections (Sec. D or E), but in this postmark the upper line touched the upper comb pattern section. It might have been cut the bottom of the upper part because it didn't fit into the upper space.

After then, the home country sent the new handstamps for the change to ordinary PO on November 21, and they switched to the new one. fig.2-1.36 shows a usage example of the postmark with "Qingdao" shifted downward. Although they had the new handstamps then, they continued using their prepared one. It seemed they didn't have a sufficient number of handstamps, so, there are some interesting usages.

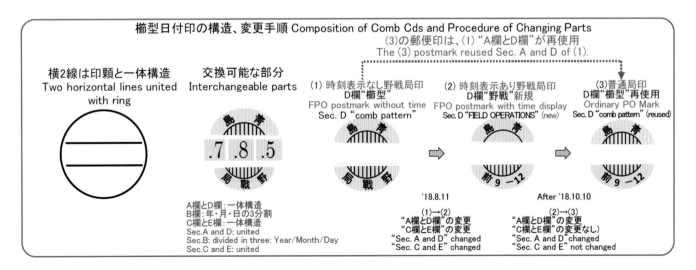

fig.2-1-34

済南局（欧文日付印）

Jinan PO (Roman Letter Cancellation)

大正7（1918）年11月21日青島、濰縣、済南に普通局が併設されると国外宛の郵便物に欧文日付印が使われました。済南局の欧文日付印の使用例を説明します。

ドイツの交換局が第一次大戦前のストラスブルグ（仏語ストラスブール）からカールスルーエに変更されました。大戦前ストラスブルグはドイツ領でしたが、大戦後にフランス領に戻りました。郵便料金が国外宛葉書4銭により桜連合葉書4銭"支那"字入りが使われました。

fig.2-1.37に大正8年9月7日済南局で受付けられたドルトムント宛の使用例を示します。fig.2-1.38が使われた郵便印です。中国内の他地域の日本局でも欧文が使われていますが、時刻表示がなく、C欄の"I.J.P.O."の活字の幅が狭い特徴があります。fig.2-1.39に逓送路を示しました。マルセイユから交換局であるカールスルーエ経由の経路です。

第一次世界大戦直後はドイツ宛の郵便はスイス、オランダなどの第三国を経由され、各国では検閲が行われました。また、地中海は騒々しく日本郵船はケープタウン経由でイギリスに航行しています。1919年1月18日から開催されたパリ講和会議で戦後の処理が決められました。fig.2-1.37の葉書は日本郵船の地中海ルートで運ばれました。

On November 21, Taisho 7 (1918), when the ordinary post offices were additionally established in Qingdao, Weixian and Jinan, they used postmarks for international mail with alphabets. I show an example of the Roman Letter cancelation of the Jinan PO.

The German exchange PO was Strasburg (Strasbourg in French) before the WWI and changed to the Karlsruhe PO. Strasburg was a part of German territory before the war and returned to France after the war. The postage was 4 sen for international postcard, and a cherry blossom UPU postcard with "China" overprint was used.

fig.2-1.37 shows a usage example to Dortmund dated on September 7, Taisho 8 (1919), accepted at the Jinan PO. fig.2-1.38 was its postmark. The Roman Letter cancellations were used in other places in China, and they were without time indication and had a narrow font style of "I.J.P.O." in the Sec. C. fig.2-1.39 shows the delivery route, which was from Marseille via the exchange PO, Karlsruhe.

Just after the outbreak of the WWI, the mails to Germany were delivered via the third country such as Switzerland or Netherland, and were censored in each country. And the Nippon Yusen navigated via Cape Town to GB because the security of the Mediterranean

欧文日付印
Roman letter Cds
使用期間 used between
'18.11.30.〜'22.10.18.

済 南 Jinan
'19.9.7
C欄 Sec. C "I.J.P.O."

fig.2-1-38

青島本局移転地 Qingdao Central PO's New Site

fig.2-1-40

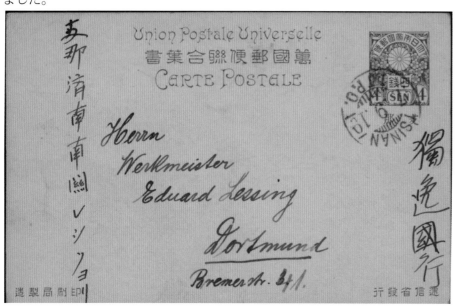

fig.2-1-37

欧州逓送鉄路
Railway Route in Europe

ドルトムント Dortmund
ストラスブール Strasbourg
カールスルーエ Karlsruhe
リヨン Lyon
マルセイユ Marseille

青字は交換局
Blue: Exchange PO
至ポートサイド to Port Said

| '19.9.7 Jinan PO | → | Jinan dep.: 9.8 8:35 | Upbound No.1 Shandong Railway | arr.: 9.8 19:16 via Qingdao PO dep.: 9.13 | Saikyo Maru Nippon Yusen | arr.: 9.15 via Moji PO dep.: 9.25 | Atsuta Maru Nippon Yusen | arr.: 11.5 Marseille dep.: 11.6 5:45 | via Strasbourg |

| arr.: 11.7 7:17 via Karlsruhe PO dep.: 11.7 10:46 | via Mannheim | arr.: 11.7 13:46 via Frankfurt dep.:11.7 14:08 | via Giessen | arr.: 11.7 21:17 Dortmund | → | Dortmund PO |

fig.2-1-39

大正 9 年 2 月 3 日所沢町に新局舎が完成し、青島本局が佐賀町から移転されました。fig.2-1.40 に移転場所を示しました。旧局舎は佐賀町支局として継続して使われました。日本租借後も中国青島廣西路市内局（郵便印の表示は青島「一」）として使われ、1949 年中華人民共和国建国後も市内局 " 廣西路郵亭 " として使われました。

青島本局（本局移転、機械式日付印）

fig.2-1.41 の封書は青島本局の移転初日の使用で郵便印は従来の印顆が継続して使われました。

大正 11 年 5 月 23 日付け青島守備軍公報で機械印の試行開始が公表されました。fig.2-1.42 に 6 月 23 日青島局で受付けられた横須賀宛の私製葉書です。青島局で機械印が使われました。この機械印は確認されている中で最初期使用です。また、無料の軍事郵便制度が廃止され有料となりました。この葉書には国内宛葉書料金 1 ½ 銭により旧大正毛紙 1 ½ 銭が使われました。この年の 12 月10 日にワシントン会議により青島守備軍の撤退と中国租借地の返還が行われました。従って、機械印の試行期間は半年程でした。日本国内宛のため逓送路は省略します。

Sea was unsafe. The Paris Peace Conference started on January 18. 1919 and set the postwar settlements. The postcard in fig.2-1.37 was delivered by the Nippon Yusen's Mediterranean route.

The Qingdao CPO's new building was completed on February 3, Taisho 9 (1920) in Tokorozawa, and it moved there from Saga. fig.2-1.40 shows the new site. The old building continued as Saga Branch. After the Japanese occupation period, this building was used for Chinese Qingdao Guangxi Road Local PO (Qingdao "1" was shown on a postmark), and after the founding of the People's Republic of China in 1949, it was used as a local post office "Guangxi Road PO."

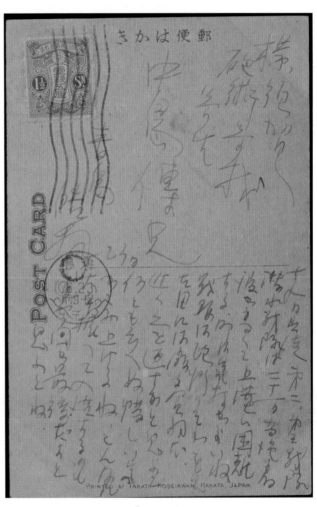

fig.2-1-42

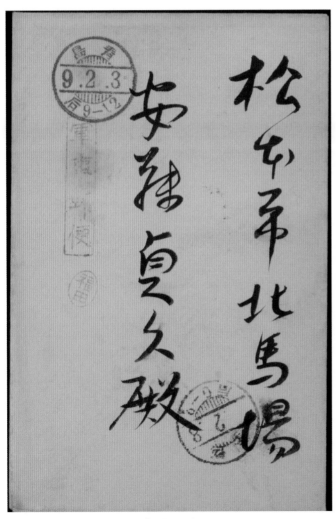

fig.2-1-41

機械式日付印
（日付拡大）
Machine-made Cds
(Enlarged date section)
使用期間 used between '22.6.23 ～ '22.11. 22

青島
11
6 .23
前9-12
〒

青島
大正11.6.23. 前9-12
最初期使用
Qingdao
'22.6.23 9-12 a.m.
Earliest Usage

fig.2-1-43

軍事郵便証票（軍事郵便が有料）

大正3年8月27日に逓信省告示第521号により始まった軍事郵便制度は、大正10年3月23日郵第961号「青島守備軍管内軍事郵便取扱制限の件」により、大正10年4月1日から無料の軍事制度が制限され郵便料金が有料となりました。そこで、守備隊の下士官、兵士に1ケ月に2枚の軍事郵便証票が支給されました。この証票は1種便（封書）に限定され、2種便（葉書）は兵士でも有料でした（fig.2-1.42が有料の例）。

山東半島で支給された軍事郵便証票は、1種便料金3銭により、旧大正毛紙切手3銭に"軍事"と加刷されました。加刷文字の"軍"と"事"の間隔が4.25mmで、大正10年5月〜大正11年12月末の使用が確認されています。

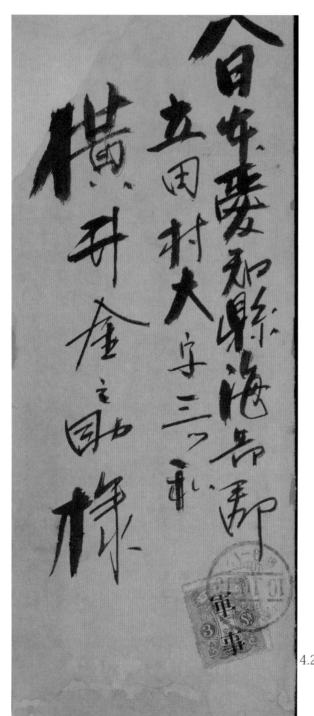

fig.2-1-44

Qingdao CPO (Move of CPO, Machine Date Cancel)

fig.2-1.41 is the first day usage of the move of the Qingdao CPO and the same postmark of the old office continued to be used.

The Qingdao Defensive Army Report dated May 23, Taisho 11 (1920) announced the start of the trial use of machine cancellation. fig.2-1.42 is a private postcard to Yokosuka, accepted at the Qingdao PO on June 23 with a machine cancellation. This machine cancellation was the earliest usage of all known by now. Then, the postage-free military mail system was abolished and all the mails became postage-paying. This postcard was postage-paying, the postage was 1 ½ sen for domestic postcard and it bears a 1 ½ sen of Taisho granite old die series. On December 10 of this year, according to the Washington Conference, the Qingdao Defensive Army withdrew and the Chinese leased territories were receded. Therefore, the machine cancellation trial only continued about half a year.

Since the usage example was to Japan, the delivery route is omitted.

Military Stamp (Postage-Paying Military Mail)

The military mail system which started on August 27, Taisho 3 (1914), according to the Ministry of Communications' Notification No.521, was limited in handling postage-free mails from April 1, Taisho 10 (1921), by the Postal Notification No. 961 "the Limitation of Handling Mails in Qingdao Defensive Army's Control Area" dated March 23, Taisho 10 (1921), and it became a postage-paying service. Then, two military labels were issued monthly to a noncommissioned officer or serviceman of the Defensive Army. This stamp was limited to use for the 1st class mail (letter), and for the 2nd class mail (postcard) even soldiers had to pay the postage (fig.2-1.42 is an example of postage-paying mail).

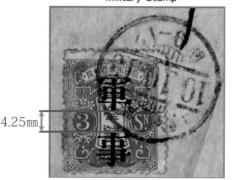

普通局印
Ordinary PO Mark
使用期間 used between '18.9.27〜
'22.12.10

軍事郵便証票拡大コピー
Military Stamp

4.25㎜

青島
大正10.10.13.
C欄"前9-12"
Qingdao
'21.10.13
Sec. C "9-12 a.m."

fig.2-1-45

尚、山東半島では、正規の軍事郵便証票が4月1日開始に届かず、現地で"軍事"と加刷した切手が配布されました。その後、正規版が届き使用されました。fig.2-1.44 の封書は大正10年10月13日に青島局で受付けられた立田宛の封書です。正規版の使用でした。

ここで、日本宛の逓送路について触れておきます。青島から日本宛の郵便物は、青島〜大阪間の定期船、若しくは、青島〜大連・大連〜下関間の定期船が利用されました。交換局は門司局と下関細江局でした。小包は全て神戸局が交換局でした。九州・沖縄宛は門司局、本州・北海道は下関細江局に送られました。

従って、二つの郵袋が定期船に積込まれました。これは、ドイツ租借時代も同様にミュンヘン局、ストラスブルグ局宛の二つの郵袋が積込まれました。それぞれに自国の交換局があり、宛先により仕分けができますが、そうでなければ郵袋は一つです。逓送経路を調べるためには交換局の特定が第一です。交換局が特定できれば、経路の特定は容易と言えます。

行徳にある郵政博物館資料センターには各国の交換局が記された地図があります。また、各種印顆の当時使われた実物もあります。事前申し込みが必要ですが、閲覧できます。但し、コピーができませんのでカメラで撮影するしかありません。カメラ撮影はOKです。事前に申し込むと丁寧に資料を探して頂けます。閲覧は平日の指定がありますので、普段行くことができず、毎年盆休みを利用して行っています。

The military labels issued in the Shandong Peninsula was for the 1st class mail which postage was 3 sen. So, they were 3-sen stamps of the Taisho granite old die series with "military" overprint. The gap between the first two characters of the overprint was 4.25mm, and the usages have been known from May in Taisho 10 (1921) to the end of December, Taisho 11 (1922).

The regular military stamps didn't arrived before the start date, April 1, and the provisional stamps with "Military" overprint were issued in the Shandong Peninsula. After then, the regular stamps arrived and were used. fig.2-1.44 is a letter to Tatsuta, accepted at the Qingdao PO on October 13, Taisho 10 (1921). It bears a regular military stamp.

I refer about the delivery route to Japan. A mail from Qingdao to Japan was delivered by a regular liner between Qingdao and Osaka, or from Qingdao to Dalian and Dalian to Shimonoseki. The exchange offices were Moji PO and Shimonoseki-Hosoe PO. All the parcels were exchanged at the Kobe PO. A mail to Kyushu or Okinawa was delivered to the Moji PO, and that to Honshu and Hokkaido to the Shimonoseki-Hosoe PO.

Therefore, a regular liner took on two mail bags. In the German occupation period, a ship also took on two mail bags for the Munich PO and the Strasburg PO. Each country had its own exchange post offices and you can know which bag had the mail you research if you see the destination. Without this system, a ship would have taken on only one mail bag. When you research the delivery route, you have to identify first the exchange PO. It will be easy to know the route if you find the exchange PO.

The Research and Documentation Centre of the Postal Museum of Japan in Gyotoku (Ichikawa city, Chiba) has a map of exchange POs worldwide. And you can find original postmarks then. You need to ask in advance, but you can see them. However, it is prohibited to copy them, and you can only take photos. Photography is permitted. If you ask in advance, they will search every document you need. You can see them only week days. In my case, I cannot visit there often and use every summer vacation to do so.

逓送路 Route

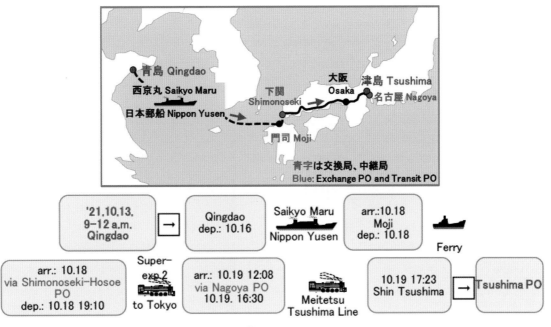

fig.2-1-46

山東鉄道車内局（鉄郵印）

　山東鉄道は大正4年（1915）1月から一般乗客の運行が開始され、同年6月21日から青島〜済南間の列車内で郵便物の引受が開始されました。

　運行当初は毎日1便が青島〜済南間を相互に発車していましたが、大正6年12月から、昼便、夜行便の毎日2便の運行になりました。

　郵便印は、表示から初期、中期、後期に分けられます。（　）内は確認されている使用期間

- 初期型印：C欄上・下便
  （大正4年6月22日〜大正5年9月13日）
- 中期型印：C欄青島局郵便係員、E欄上・下便
  （大正6年2月24日〜大正6年10月9日）
- 後期型印：C欄上一・二、下一・二便
  （大正7年4月25日〜大正11年11月30日）

　青島が起点で下り、上り、一は一番列車を指します。

　中期型印のみC欄に「青島局郵便係員」D欄に上・下の郵便印が大正6年のみ使用されました。

　この表示については資料を探しましたが、中国局が関係していると考えていますが、未だに見つかっていません。後程説明する済南局"D欄なし"も使用の経緯がわからない郵便印です。

　fig.2-1.48に中期型印の使用を、fig.2-1.49に鉄郵印を示しました。

Shandong Railway Post Office (Railway Postmark)

The Shandong Railway started the service for the public in January, Taisho 4 (1915), and on June 21 of the same year, the train between Qingdao and Jinan started handling mails.

Initially, it run one service daily from each side of Qingdao and Jinan, and from December, Taisho 6 (1917), changed to two services daily, day and night.

The postmarks were classified by description, into early, mid and late period. () is the known use period

- Early Period Postmark: "upbound" or "downbound" in the Sec. C
  (June 22, Taisho 4 [1915] to September 13, Taisho 5 [1916])
- Mid Period Postmark: "Qingdao PO Officer" in the Sec. C and "upbound" or "downbound" in the Sec. E
  (February 24 to October 9, Taisho 6 [1917])
- Late Period Postmark: "upbound No.1 or 2" or "downbound No.1 or 2" in the Sec. C
  (April 25, Taisho 7 [1918] to November 30, Taisho 11 [1922])

The starting place of up and down bound was Qingdao, and "1" meant the first train.

The mid postmark was the only one which had "Qingdao PO Officer" in the Sec. C and "upbound" or "downbound" in the Sec. E. It was used just in Taisho 6 (1917).

I believe there was a certain connection between this characteristic description and the Chinese post offices, but I haven't found any document to clarify it yet. The postmark of the Jinan PO without Sec. D, which will be described later, is also a postmark which history was unknown.

fig.2-1.48 shows a usage example of the mid period postmark and fig.2-1.49 shows a railway postmark.

山東鉄道主要駅(沿線開局地)
Shandong Railway Main Stations (with Postal Service)

fig.2-1-47

鉄郵印(中期型)
Train Postmark (Middle Period Type)
使用期間 used between
'17.2.24 〜'17.10.9

青島済南間
大正6.8.30.
A欄"青島済南間"
C欄"青島局郵便係員"
E欄"上便"
Qingdao-Jinan
17.8.30
Sec. A "Qingdao-Jinan"
Sec. C "Qingdao PO Officer"
Sec. E "Upbound"

fig.2-1-49

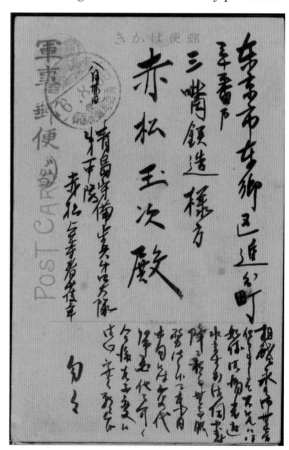

fig.2-1-48

青島〜大阪間定期船には郵便函が設置され、投函された郵便物は下関局（当時下関には二つの郵便局がありました。）で舩函揚の角印が押印されました。この定期船内の郵便函は日本軍撤退後も継続されました。

fig.2-1.50 は大正 8 年 3 月 11 日に青島を出港した薩摩丸船の郵便函に投函され、下関細江局で大正 8 年 3 月 13 日の日付印と舩函揚印が押印されました。当時の郵便結束図に列車の指定があり、下関発 19:10 の東京行の特急列車で大垣に運ばれました。郵便料金は国内宛封書 3 銭により旧大正毛紙 1 ½ 銭 " 支那 " 加刷 2 枚が貼られました。

下関は大連間、天津間などに定期船があり、船内で投函された郵便物には、いずれもこの船函揚の押印があります。差出人の確認と下関の日付印から当日入港した船舶の確認が必要になります。

A mailbox was established in a regular liner between Qingdao and Osaka, and the mails posted there were handstamped with a "mail by ship mailbox"rectangle mark at the Shimonoseki PO (There were two post offices in Shimonoseki then). The mailbox in a regular liner continued to be used after the evacuation of Japanese forces.

fig.2-1.50 is a letter posted in a mailbox in Satsuma Maru which left Qingdao on March 11, Taisho 8 (1919), and the Shimonoseki-Hosoe PO cancelled on March 13, Taisho 8 (1919) with a "Paquebot" mark. The postal service network map then showed the train designation, and you will find by that, it was delivered to Ogaki by a super-express train of 19:10 from Shimonoseki to Tokyo. The postage was 3 sen for domestic mail, and it bears two 1 ½ -sen stamps of the Taisho granite old die with "China" overprint.

Shimonoseki had a regular liner service with Dalian or Tianjin, and all the mails posted on ship had a "Paquebot" mark. To find the route, you have to check the sender, and the ships arrived on that day by the date cancellation of Shimonoseki.

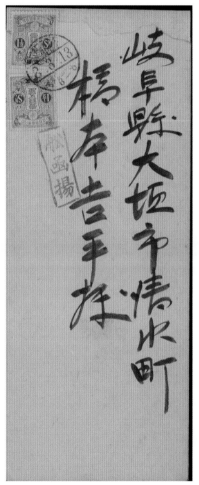

fig.2-1-50

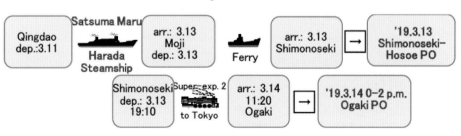

fig.2-1-51

済南 "D 欄なし" 普通局印

「山東日支通信連絡細項取極め」により青島、濰縣、済南局が中国との交換局となりました。大正7年10月~11月の間に青島局、濰縣局、済南局ではD欄"野戦"から"櫛型"の郵便印に変更されました。

済南局で翌年の大正8年6月から3ケ月間 "D 欄なし" の郵便印が使われました。

Fig.2-1.52に大正8年7月19日に済南局で受付けられた長野県上伊那郡宛の使用を示しました。郵便印はfig.1-2.53に示します。なぜこの郵便印が使われたかはわかっていません。但し、普通局になり "D 欄櫛型" の郵便印になってから櫛の縦線がないもの、櫛の線が太い郵便印が使われています。印顆の不足があり、このような郵便印が使われたのかもしれません。尚、この郵便印が使われた期間に "D 欄櫛型" の普通局郵便印の使用は確認されていません。

日本国内でも昭和21年に同じ "D 欄なし" の郵便が使われています。Fig.2-1.54に本郷から静岡県宛の楠公葉書5銭の郵便印の部分を示しました。国内の櫛型印の時期に "D 欄なし" は少ないと思います。戦争が激化した時期ですので、材料不足との理由でしょうか。

Jinan Ordinary Postmark "without Sec. D"

According to the "Agreement in Detail of Correspondence between China and Japan in the Shandong Province," the Qingdao, Weixian and Jinan POs became the exchange POs with China. From October to November in Taisho 7 (1918), at the Qingdao, Weixian and Jinan POs, the postmarks were changed from "filed operations" to "comb pattern" in the Sec. D.

At the Jinan PO, a postmark without "Sec. D" had been used in three months from June of the next year, Taisho 8 (1919).

Fig.2-1.52 shows a usage example to Ina-gun in Nagano, accepted at the Jinan PO on July 19, Taisho 8 (1919). You can see the postmark in fig.1-2.53. It is unknown why this postmark was used. However, after the post offices changed to ordinary ones and the postmarks also changed to those with "comb pattern in Sec. D", they used sometimes the postmarks with the Sec. D without vertical lines or with border lines. It would be possible that these postmarks were used because the regular handstamps were not sufficient for the demand. In the period when this postmark was used, no usage of ordinary postmark with "comb pattern in Sec. D" has been known.

In Japan, a postmark "without Sec. D" was also used in Showa 21 (1946). fig.2-1.54 is a postmark on a 5-sen NANKO postcard sent from Hongo. There should have been a small number of "without Sec. D" in the period of comb style postmark. Since it was the same period when the war heated up, it would have been used by lack of materials.

野戦局から軍事局

普通局になった濰縣、済南を除く租借地外の山東鉄道沿線各局は、大正9年1月29日に野戦局から軍事局に改称されました。改称された沿線局と後述する秘密局の開局地をfig.2-1.55に示しました。

これらの郵便局は①から④の変遷でした。

① 時刻表示なし野戦局印
② 時刻表示あり野戦局印
③ 軍事局暫定印
④ 軍事局印

郵便印の変遷をfig.2-1.56にまとめました。各郵便印の日付は確認されている最初期です。

Field PO to Military PO

The Shandong Railway zone post offices out of leased territories, but Weixian and Jinan which became ordinary POs, changed their name from Field PO to Military PO on January 29, Taisho 9 (1920). The name-changed post offices and secret post offices which will be mentioned later are shown in fig.2-1.55.

These post offices changed the postmarks (1) to (4).

(1) Field Postmark without Time
(2) Field Postmark with Time
(3) Military Provisional Postmark
(4) Military Postmark

fig.2-1.56 shows the changes of postmarks. The date on each postmark is the earliest use known by now.

C欄、D欄の表示：年月日は確認されている変更初日を例として記載
Sec. C and D: The written date is the recorded first changed day

fig.2-1-56

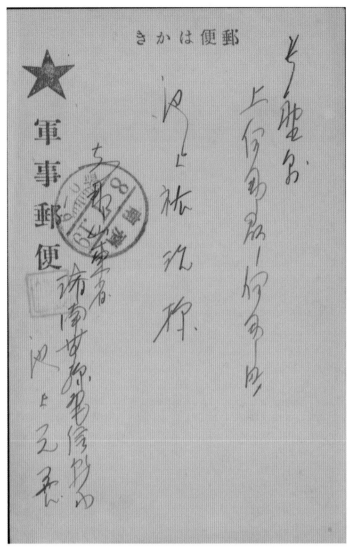

fig.2-1-52

普通局印
D欄"櫛型"なし
Ordinary PO Postmark
Sec. D without "comb pattern"
使用期間 used between
'19.6.3～'19.9.3

済　南
大正8.7.19.
C欄"前0-9"
Jinan
'19.7.19
Sec. C "0-9 a.m."

fig.2-1-53

fig.2-1-54

山東鉄道沿線開局地 Shandong Railway Zone with Postal Service

fig.2-1-55

Fig.2-1.57 に膠州局で大正 5 年 10 月 18 日に受け付けられたコペンハーゲン宛の私製葉書を示します。膠州局が①時刻表示なしの野戦局時期の使用例です。郵便料金は国外宛葉書 4 銭により、旧大正毛紙 1 銭 +3 銭が私製絵葉書に貼られました。門司局の中継印、コペンハーゲン局の着印があります。

Fig.2-1.58 に膠州日付印（再現）、門司中継印、コペンハーゲン着印を示しました。

この葉書も日本、アメリカを経由して、大西洋を渡り、ロッテルダムからコペンハーゲンに逓送されました。1 ヶ月半の日数を要しました。

大西洋航路は、NEW YORK TIMES に汽船の出入港が掲載されています。参考に 1916 年 11 月 14 日の出港掲載を fig.2-1.59 に示しました。国会図書館で閲覧しました。国会図書館には、各国の新聞があり、重宝しています。コピーも安価ですので、調査するのにお勧めです。

この時期の太平洋航路は日本の汽船会社の定期航路が増加し、大西洋航路のドイツ、北欧宛はオランダ、ベルギーの汽船会社が主流となっています。ブレーメン、ハンブルク出入港のドイツ系汽船会社の広告は掲載されていません。戦争の影響です。

Fig.2-1.57 shows a private postcard to Copenhagen, accepted at the Jiaozhou PO on October 18, Taisho 5 (1916). It is a usage example of (1) field postmark without time at the Jiaozhou PO. The postage was 4 sen for international postcard, and the private postcard bears a 1-sen and a 3-sen stamps of Taisho granite old die series. It had a transit mark of the Moji PO and an arrival mark of Copenhagen.

fig.2-1.58 shows a Jiaozhou date cancel (reproduction), a Moji transit mark and a Copenhagen arrival mark.

This postcard was also delivered via Japan and U.S., across by the Atlantic Ocean to Rotterdam, and to Copenhagen. It took a month and a half.

As for the route across the Atlantic, the arrival and departure information of the steamships were notified in NEW YORK TIMES. fig.2-1.59 is the notification of departure dated November 14, 1916 for reference. I found it in the National Diet Library. In the Library, there are worldwide newspapers and I appreciate it. You can copy the documents at a low price and I recommend there as a location to use for research.

Japanese steamship companies' regular ships increased in the Pacific route then, and the main countries handling mails were Germany in the Atlantic route, and the Netherland and Belgium to the Nordic countries. There was not a notification of German steamship companies which arrived at and left from Bremen and Hamburg. It was because of the war.

筆氏聖梅村松 花の粉白 （品出會覧展術美書部文回七第）

fig.2-1-57

時刻表示なし野戦局
FPO postmark without time
使用期間 used between
'15.4.9～'17.2.15

膠 州 Jiaozhou FPO
大正5 ('16).10.18

中継印コピー
Transit Mark

門 司 Moji '16.10.23

fig.2-1-58

着印コピー
Arrival Mark

コペンハーゲン Copenhagen
'16.12.1 4-5ε
eftermiddag= p.m.

NYタイムズ THE NEW YORK TIMES 1916.11.14

SAIL TUESDAY.

Noorderdijk, Rotterdam	——	12:00 M.
Palumo, Genoa	8:30 A.M.	12:00 M.
Alicanti, Cadiz	8:30 A.M.	12:00 M.
Lenape, Jacksonville	——	12:00 M.

fig.2-1-59

青州局の時刻表示あり野戦局

青州局時刻表示あり野戦局時期の使用を fig.2-1.61 に示しました。鹿児島県加治木局で大正8年7月31日に受付けられた青州宛で、8月5日の着印があります。郵便料金は国内宛封書3銭と書留7銭で10銭でした。旧大正毛紙10銭の切手が使われました。郵便印の再現と逓送路を fig.2-1.62 と fig.2-1.63 に示しました。

Field Postmark with Time at the Qingzhou PO

fig.2-1.61 shows a postmark with time at the Qingzhou PO in the field PO period. It was a letter to Qingzhou, accepted at the Kajiki PO in Kagoshima on July 31, Taisho 8 (1919). It had an arrival mark dated August 5. The postage was 10 sen in total consisting of 3 sen for domestic letter and 7 sen for registered mail. A 10 sen Taisho granite old die series was used. fig.2-1.62 shows a reproduced postmark and fig.2-1.63 shows the delivery route.

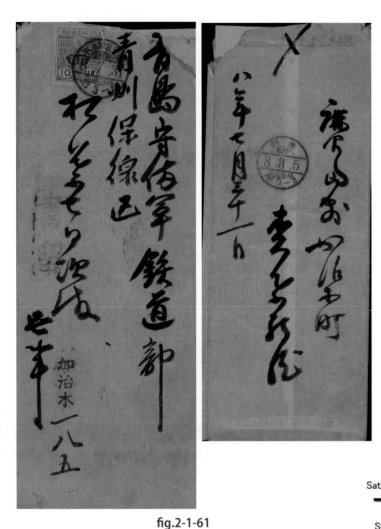

fig.2-1-61

時刻表示あり野戦局印
FPO postmark with time display
使用期間 used between '18.9.5～'19.11.4

青 州/大正8.8.5.
C欄"后0-3"
D欄"野戦"
Qingzhou FPO/ '19.8.5
Sec. C "0-3 p.m."
Sec. D "FIELD OPERATIONS"

fig.2-1-62

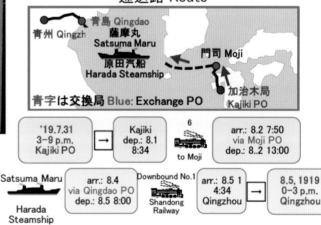

fig.2-1-63

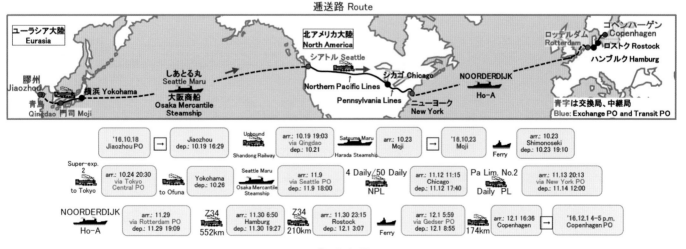

fig.2-1-60

軍事局暫定印と軍事局印

軍事局に移行過程に現地で暫定が調製されました。高密局と坊子局の使用例をまとめて紹介します。

大正9年1月29日に野戦郵便局から軍事郵便局に改称され郵便印のD欄表示が"軍事"に変更されました。但し、本国から変更の印顆が届くまで、各局でD欄"櫛型"の暫定印が調製されました。fig.2-1.64は高密局で大正9年2月9日に受付けられた青島守備軍宛の軍事郵便です。本国から未だ新印顆"D欄軍事"が届かず、現場で調製され使われました。fig.2-1.65は坊子局で大正9年2月26日に受付けられた長野県下伊那郡宛の軍事郵便で、新印顆の郵便印です。

暫定印と軍事局印の再現をfig.2-1.66に示しました。左の高密の郵便印は租借地の普通局の郵便印と同じ表示です。

この野戦局から軍事局への名称の改称は、郵便印の変更からわかる通り、逓信省に事前の周知のないままに行われた、青島守備軍の独断専行と考えられます。軍事郵便という名称もこの時初めて使われました。野戦郵便局制度に準ずると青島守備軍公報に記載されています。準備不足を象徴しています。戦闘が終結しているのに野戦局とはおかしいとの中国や諸外国からの指摘をかわす目的でした。各葉書の逓送路は省略します。

Military Provisional Postmark and Military Postmark

When the post offices were changing to military post offices, provisional postmarks were prepared. Here I introduce two usage examples of the Gaomi PO and the Fangzi PO.

On January 29, Taisho 9 (1920), field PO changed the name to military PO and the Sec. D of a postmark changed to "military." However, until the arrival of the new regular handstamps from the home country, each post office prepared a provisional handstamp with "comb pattern" in the Sec. D. fig.2-1.64 is a military mail to the Qingdao Defensive Army, accepted at the Gaomi PO on February 9, Taisho 9 (1920). The new regular handstamp with "Military in the Sec. D" didn't arrived from the home, the post office prepared a provisional one. fig.2-1.65 is also a military mail to Shimoina-gun in Nagano, accepted at the Fangzi PO on February 26, Taisho 9 (1920) , with a postmark of new regular handstamp.

fig.2-1.66 shows the provisional and military reproduced postmarks. The Gaomi postmark on the left was the same of the ordinary post offices in the leased territories.

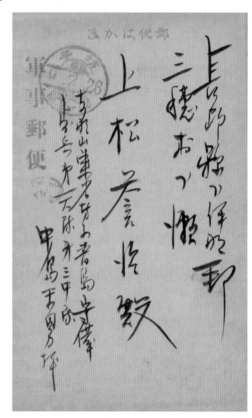

fig.2-1-65

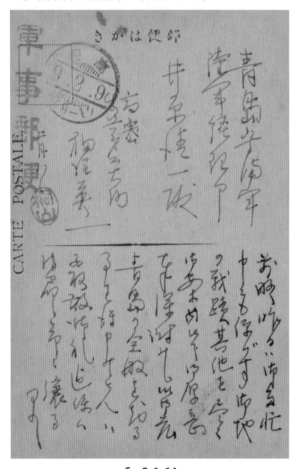

fig.2-1-64

軍事局印(暫定印) MPO Postmark (Provisional) 使用期間 used between '20.2.9〜 '20.4.4	軍事局印 MPO Postmark 使用期間 used between '20. 2.26 〜 '22.8.6
高密 Gaomi 大正9('20).2.9. 最初期使用 C欄"前9-12"、D欄"櫛型" Earliest Usage Sec. C "9-12a.m.", Sec. D "comb pattern"	坊子 Fangzi 大正9('20).2.26. 最初期使用 C欄"前9-12"、D欄"軍事" Earliest Usage Sec. C "9-12a.m.", Sec. D "MILITARY"

fig.2-1-66

張店第三（周村）：秘密局

　中国に対して開局を極秘にするため、周村に張店第三の名称で大正8年4月16日に開局されました。秘密局は租借外の山東鉄道停車駅3箇所に開局されました。

　fig.2-1.67 に大正10年1月1日付けの玉島（岡山県）宛の葉書を示します。年賀の葉書で、日本の年賀特別取扱い（元日の日付を押印）が青島でも実施されていましたので、少し早く周村局で受付けられました。この葉書は無料の軍事郵便扱いでした。

　この年賀の特別取扱いは日本国内で1899年から一部の郵便局で開始され、1905年には全国の郵便局で実施されました。青島局でも大正6年からの使用例を確認しています。再現した郵便印を fig.2-1.68 に示しました。

　年賀の特別取扱いは、ずいぶん古くからの制度だったことがわかります。最近は年賀状に郵便印が押されなくなりました。一昨年東京オリンピック寄付金付きの年賀状が発売され、購入しました。記念に自分宛にも差出したのですが、元旦に配達されましたが日付印が押印されていないことに気が付き、近所の深川郵便局に行って押印して頂きました。元日の日付印の押印が行われなくなるのは残念です。

The name change from field PO to military PO is thought, as you imagine by the situation of postmark, to have been carried out arbitrarily on the Qingdao Defensive Army's own authority, without making it known to the Ministry of Communications. The name of Military Post was used for the first time. The Qingdao Defensive Army Report said it was equivalent to the field PO system. Typical description for poor preparation. It was to avoid a comment of China or other countries that the name of field PO was not suitable because the battle was over. The delivery routes of each mail are omitted.

Zhangdian 3 (Zhoucun): Secret PO

It was established on April 16, Taisho 8 (1919) in Zhoucun with the name of Zhangdian 3 to hide it from China. There were three secret post offices established in the Shandong Railway Stations out of leased territories.

fig.2-1.67 shows a postcard to Tamashima (Okayama) dated January 1, Taisho 10 (1921). It was a New Year's card. Since in Qingdao it was also available to handle New Year cards as a special postcard for Japanese New Year (with the postmark of New Year's Day), this postcard was accepted a little before the day of Jan. 1st at the Zhoucun PO. It was handled as postage-free military mail.

The special handling for New Year's card started in Japan in 1899 just at some post offices, and since 1905, it became available at post offices throughout the country. The usages at the Qingdao PO have been known from Taisho 6 (1917). fig.2-1.68 shows the reproduced postmark.

Now you know the special handling for New Year's card is quite a long-standing system. Recently, we haven't had postmarks on New Year's cards. The New Year postcards with a donation to the Tokyo Olympics sold two years ago and I bought them. I sent one to myself as a memento. I received it on the New Year's Day, but it didn't have a date cancellation. So, I asked my neighbor Fukagawa PO to handstamp by the New Year's Day cancel on it. It is a pity that New Year Greeting card is delivered without cancellation of Jan. 1st.

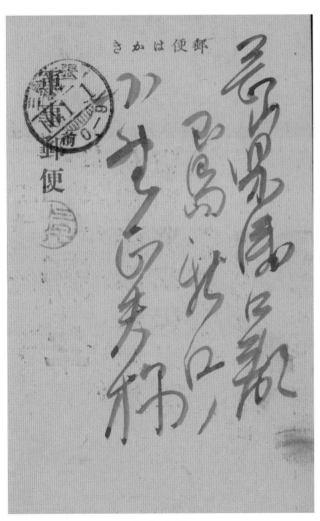

fig.2-1-67

軍事局印
MPO Postmark
使用期間 used between
'20.10.72 ～ '21.7.2

張店第三　大正10.1.1.
C欄"前0-9"　D欄"軍事"
3rd Zhangdian '21.1.1
Sec. C "0-9a.m." Sec. D "MILITARY"

fig.2-1-68

淄川炭鉱局：軍事局時期   Zichuan Mine PO: Military PO Period

　無料の軍事郵便が大正 10 年 4 月 1 日より有料となりました。軍事郵便証票は配布されましたが 1 種便（封筒）に限定されましたので、葉書は有料でした。fig.2-1.69 は青島守備軍職員が淄川炭鉱局に差出した日本宛の葉書で有料となりました。旧大正毛紙 1 ½ 銭の切手が貼ってあります。

　この淄川炭鉱局は鉄山局から名称が改称された郵便局です。fig.2-1.70 に開局地を示しました。山東鉄道の張店駅から博山に博山支線があります。その支線の途中の淄川駅から淄川炭鉱に分線がありました。博山にも秘密局の張店第二が開局しました。また、山東鉄道金嶺鎮駅の北に張店鉄山局も開局されました。このあたり一帯は石炭が産出され、賑わっていたようです。逓送路は省略します。

The military mails changed from postage-free to postage-paying on April 1, Taisho 10 (1921). The military labels were still issued but just for 1st class mail (letter), so they had to pay for sending postcards. fig.2-1.69 shows a postage-paying postcard sent by an officer of the Qingdao Defensive Army from the Zichuan Mine PO to Japan. It bears a 1 ½-sen Taisho granite old die series.

The Zichuan Mine PO was the ex-Iron Mine PO. fig.2-1.70 shows the location of post offices. The line between Zhangdian and Boshan in the Shandong Railway was the Boshan Local Route. And between a way station of the local route, the Zichuan Station and the Zichuan Mine, there was a branch line. Also, in Boshan, a secret PO of Zhangdian 2 was established. And north of the Jinlingzhen Station in the Shandong Railway, the Zhangdian Iron Mine PO was established. It seemed that this area was a busy place based on mining. The delivery route is omitted.

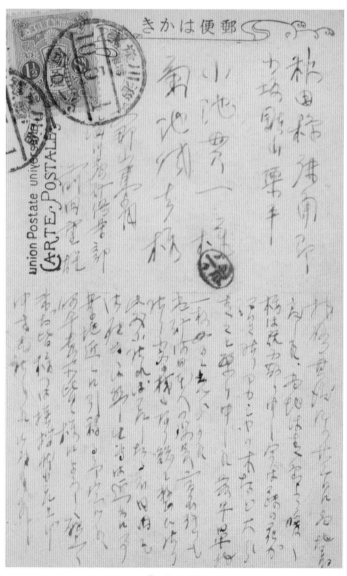

fig.2-1-69

淄川炭鉱誌(大正6年7月)を基に作図
Map based on Zichuan Mine Report
(Jul. 1917)

fig.2-1-70

青字は大正8年以降の開局
Blue: established after 1919

fig.2-1-71

青島市内局

大正 8 年から青島市内に 5 局が新たに開局されました。青島市内開局地を fig.2-1.71 の図に示しました。図の下に距離を示しています。僅か 2km 四方の狭い範囲に 9 局が開局しました。郵便目的とは考えられません。主目的は為替業務と考えています。為替印の使用は確認例が多いのですが郵便の実逓便が少ないことがその理由です。

Qingdao Local PO

After Taisho 8 (1919), five post offices were newly established in Qingdao city. fig.2-1.71 shows Qingdao local post offices. It has a scale in the bottom. As many as nine post offices were established in an area 2km square. It must not have been for postal service. I suppose it was mainly for foreign exchange operations. It is because there are so many usages of postmarks on money order but a few actually sent mails.

青島天津町局

青島市内局の国外宛の使用例を二つ紹介します。青島天津町局は大正 8 年 6 月 16 日に青島本局の近傍に開局されました。fig.2-1.72 の封書は青島天津町局で大正 9（1920）年 6 月 18 日に受付けられたハンブルク宛です。日本、アメリカを経由し、ニューヨークから大西洋を横断してオランダ経由でドイツに送られました。

大西洋航路によるドイツ・北欧宛は、第一次大戦前はドイツ系の汽船会社が活躍していましたが、大戦後はオランダ（Ho-A）、ベルギー（Red Star Line）の汽船会社が主体となりました。青島天津町局の実逓便は 3 例のみの確認です。郵便料金は国外宛封書 10 銭と書留 10 銭で 20 銭となり旧大正毛紙 10 銭 "支那" 加刷が 2 枚貼られています。

Qingdao-Tianjin PO

Here I show two usage examples of Qingdao local post offices. The Qingdao-Tianjin PO was established on June 16, Taisho 8 (1919) near to the Qingdao CPO. fig.2-1.72 is a letter to Hamburg, accepted at the Qingdao-Tianjin PO on June18, Taisho 9 (1920). It was passed by Japan and U.S., and delivered from New York, across over the Atlantic, via Netherland, to Germany.

The mails to Germany or the Nordic countries across the Atlantic were delivered by German steamship companies before the WWI, but after the war, the steamships of the Netherland (Ho-A) and Belgium (Red Star Line) mainly worked for that. Only three usage examples of the Qingdao-Tianjin PO have been known. The postage was 20 sen consisting of 10 sen for international letter and 10 sen for registered mail, and it bears two 10-sen stamps of the Taisho granite old die series with "China" overprint.

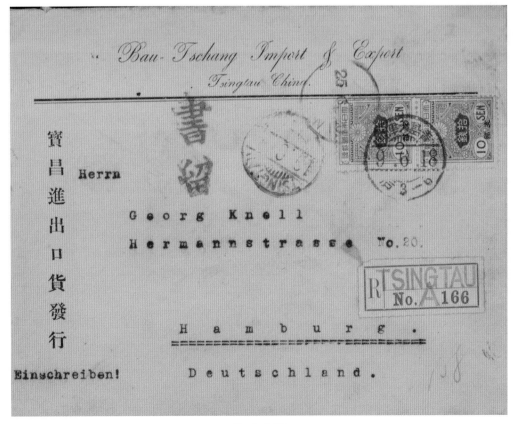

fig.2-1-72

この封書の裏面に押印されている郵便印と横浜局の中継印（再現）を並べるとfig.2-1.73のようになります。横浜局ではゴム印が使われました。ゴム印は新規の状態では円形ですが、経年劣化が進むと左右が跳びでたようになるのが特徴です。その様子も再現してみました。

書留シールに"A"と押印されています。これは青島天津町局を示します。"B"青島大鮑島、"D"青島佐賀町が確認されています。

The postmarks and the (reproduced) transit mark of Yokohama on the back are shown in fig.2-1.73. Rubber handstamps were used in Yokohama. It was a characteristic of rubber handstamp that a new one shaped circle but the both side lines jutted out by aging. I reproduced it, too.

The character "A" was on the registered label, which indicates Qingdao-Tianjin PO. "B" is Qingdao-Dabaodao and "D" is Qingdao-Saga.

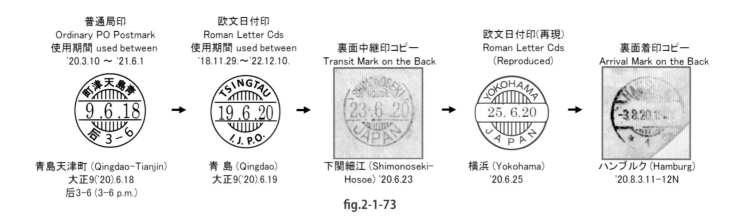

fig.2-1-73

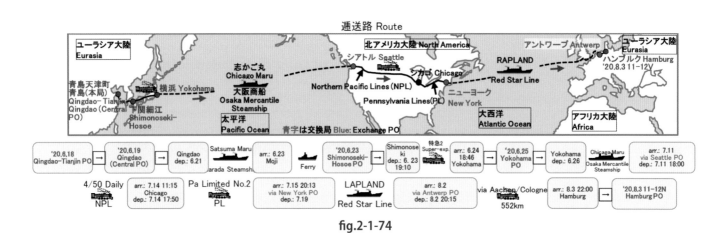

fig.2-1-74

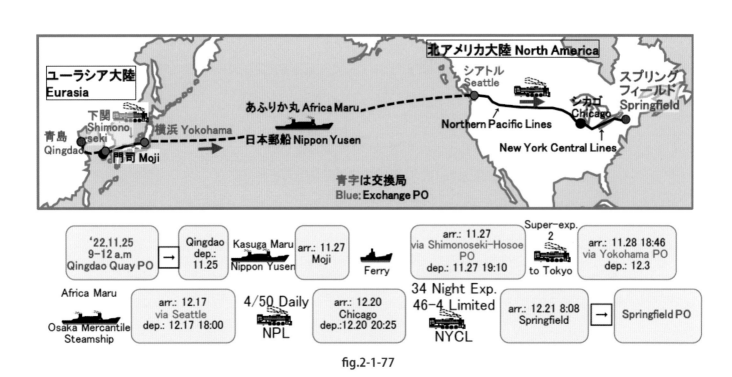

fig.2-1-77

青島野戦郵便局埠頭出張所は、市内局としては最も早く大正 7 年 11 月 2 日に開設され、大正 8 年 12 月 1 日普通局へ昇格しました。fig.2-1.75 に青島埠頭局の普通局時期の使用例を示します。

大正 11 年 11 月 25 日に受付けられたスプリングフィールド（アメリカ：MA）宛の封書です。郵便料金は大正 11 年 1 月 1 日に国外宛の料金が改正され、封書 20g までが 20 銭になりました。重量を軽くするために小型の封書を用いたのでしょう。旧大正毛紙 20 銭 “支那” 加刷切手が貼ってあります。

郵便印は普通局の C 欄：時刻、D 欄：櫛型の表示が使われています。この年の 12 月 10 日に青島守備軍は中国から撤退しましたので、撤退一月前の使用例です。

fig.2-1.77 に逓送路を示しました。横浜出入港の汽船を調べるには横浜の英字新聞 “The Japan Times” が便利です。出入港のみならず、シアトル、サンフランシスコ、マルセイユの入港も日付を追っかけると記載されています。国会図書館で閲覧できます。

The Quay Branch of Qingdao Field PO was the first local office which was established on November 2, Taisho 7 (1918), and it was promoted to the ordinary post office status on December 1, Taisho 8 (1919). fig.2-1.75 shows a usage example of the Qingdao Quay PO in ordinary post office status.

It was a letter to Springfield (U.S.: MA), accepted on November 25, Taisho 11 (1922). The postage for international mail changed on January 1, Taisho 11 (1922) and 20 sen was for a letter up to 20g. A small envelope might have been used to weigh lighter. It bears a 20-sen stamp of the Taisho granite old die series with "China" overprint.

The postmark was ordinary post office one: time in the Sec. C and comb pattern in the Sec. D. The Qingdao Defensive Army evacuated China on December 10 of this year, so, this is a usage of the previous month of evacuation.

fig.2-1.77 shows the delivery route. It is convenient to use "The Japan Times", an English newspaper in Yokohama to research the steamships which arrived at and departed Yokohama. It has not only the information of arrival or departure of Yokohama, but the arrival date of Seattle, San Francisco and Marseille if you track them. You can find it in the Diet Library.

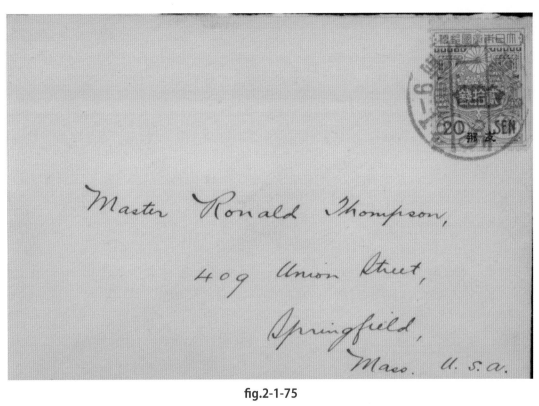

fig.2-1-75

普通局印
Ordinary PO Mark
使用期間 used between
'19.12.1〜 '22.12.10

青島埠頭
大正11.11.25.
C欄"前9-12"
Qingdao-Quay
'22.11.25
Sec.C "9-12 a.m."

fig.2-1-76

　　租借地内開局地と開局年月日を fig.2-1.78 に示しました。青島滄口局は開局当初は出張所でしたが、大正 8 年 12 月 1 日に普通局に昇格しました。

　　fig.2-1.79 は大正 4 年 6 月 17 日に李村局で受付けられ、宛先の長野東穂高局に 6 月 24 日に届いています。現在と遜色ない速さと思います。東穂高局の着印として為替印を使用しているのが面白いと思います。"むし"とは、この局の為替記号です。

fig.2-1.78 shows the post offices in the leased territories and their establishment date. The Qingdao-Cangkou PO was initially a branch and promoted to ordinary post office on December 1, Taisho 8 (1919).

fig.2-1.79 was accepted at the Li Cun PO on June 17, Taisho 4 (1915), and arrived at the addressee, Nagano Higashi Hotaka PO on June 24. The delivery speed compares favorably with now. Curiously enough, it had a money order mark as an arrival mark at the Higashi Hotaka PO. "mushi" was a mark of money order of this post office.

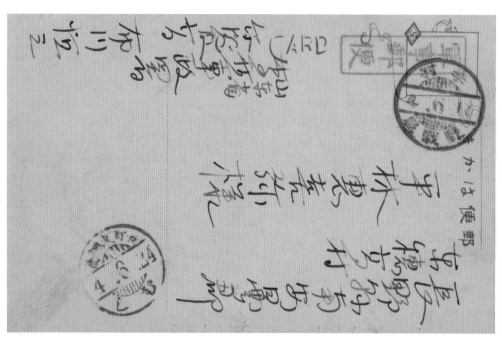

fig.2-1-79

租借地内局開局地
PO in the Leased Territory

膠州駅 Jiaozhou st.   山東鉄道 Shandong Railway

滄口 Cangkou

李村 Li Cun   租借地 leased territory

四方 Syfang

青島駅 Qingdao St.

時刻表示なし野戦局印
FPO Postmark without Time
使用期間 used between
'15.6.17 ～ '18.5 26.

4.6.17

李 村
大正4.6.17.
Li Cun
'15.6.17

着印コピー
Arrival Mark

4.6.24

長野東穂高
大正4.6.24.
むし(為替印)
Nagano Higashi Hotaka
'15.6.24
MUSHI (Postmark on the Money Order)

fig.2-1-80

租借地内局開局年月日 Open Date of PO in Leased Territory

局 名 PO Name	開局年月日 Establishment	普通局(併設) Ordinary PO (Additional Office)
[1] 四方 Syfang	'16.6.1	'18.11.21
[2] 滄口(出張所) Cangkou (local PO)	'19.3.1	'20. 5.16
[3] 李村 Li Cun	'15.4.1	'18.11.21

fig.2-1-78

青島滄口局   Qingdao-Cangkou PO

青島滄口局の使用例を fig.2-1.81 に示しました。大正 8 年 10 月 27 日に受付けられた青島宛の封書です。出張所時期の使用です。郵便料金は国内宛封書 3 銭により旧大正毛紙 3 銭 " 支那 " 加刷切手が使われました。

済南 8:55 発上一便の列車に滄口で積込まれ青島に運ばれました。受付印、着印が時刻表示ですので列車を特定するのに助かります。ドイツ租借時代もドイツ国内は時刻表示でしたので列車の特定に役立ちました。

滄口から青島までの距離は 18km 程度と近いことから、日本租借時代の後半から郵便物運搬に自動車が使われたと青島軍政史に記載されています。滄口には軍が駐留し、後年青島飛行場ができた場所です。

滄口の出張所時期には A 欄には " 青島滄口 " と表示されていましたが普通局になると " 滄口 " と変更されました。滄口局の使用例が少ないので使用期間を開局期間としています。

fig.2-1.81 shows a usage example of the Qingdao-Cangkou PO. It was a letter to Qingdao accepted on October 27, Taisho 8 (1919). The Cangkou PO was a branch then. The postage was 3 sen for domestic letter and it bears a 3-sen stamp of the Taisho granite old die series with "China" overprint.

It was delivered from Cangkou by the No1. upbound train leaving Jinan at 8: 55. Its accepted mark and arrival mark with time helped me to identify the train. In the German occupation period, the postmarks with time were also used in Germany and the time description was useful to identify the train.

Since there was a short distance of about 18km between Cangkou and Qingdao, the Qingdao Military History said they used cars to delivery mails in the latter half of Japan leased period. The forces were stationed in Cangkou, where the Qingdao Airport was constructed later.

When the Cangkou PO was a branch, the postmark had in Sec. A "Qingdao-Cangkou" and when it was promoted to the ordinary post office, it changed to "Cangkou." There are not so many examples of the Cangkou PO. The recorded usage period is almost same as the opening period of the post office.

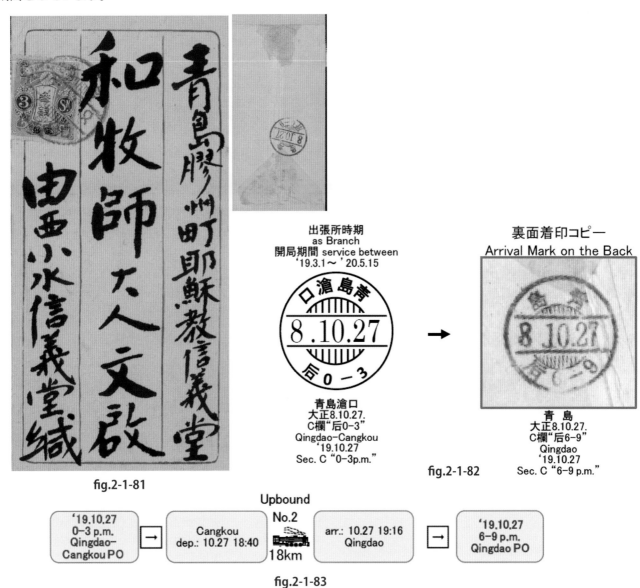

fig.2-1-81

出張所時期
as Branch
開局期間 service between
'19.3.1〜'20.5.15

青島滄口
大正8.10.27.
C欄"后0-3"
Qingdao-Cangkou
'19.10.27
Sec. C "0-3p.m."

裏面着印コピー
Arrival Mark on the Back

青 島
大正8.10.27.
C欄"后6-9"
Qingdao
'19.10.27
Sec. C "6-9 p.m."

fig.2-1-82

Upbound
No.2
18km

'19.10.27
0-3 p.m.
Qingdao-
Cangkou PO → Cangkou
dep.: 10.27 18:40   arr.: 10.27 19:16
Qingdao → '19.10.27
6-9 p.m.
Qingdao PO

fig.2-1-83

2 中華郵政

2 Republic of China Postal Service

日本租借期間中も青島を除く地域では中華郵政による郵便の集配は行われています。fig.2-2.1 に民国 9 年郵便結束図を基に作図したものを示しました。大正 7 年 10 月 10 日の日本との「山東日支通信連絡細項取極め」により日独戦争で閉鎖されていた中国青島局開局と山東鉄道の中国局の利用が可能となりました。結束図には山東鉄道を利用した逓送路が示されています。尚、赤字は次の使用例の局を示しています。

During the Japan leased period, in all the places through China but Qingdao, the Republic of China Postal Service collected and delivered mails. fig.2-2.1 shows a map based on the Postal Service Network of Minguo 9 (1920). According to the "Agreement in Detail of Correspondence between China and Japan in the Shandong Province" with Japan on October 10, Taisho 7 (1918), the Chinese Qingdao PO which had closed during the Japanese-German War and the Chinese post offices in the Shandong Railway zone were able to be used. The network map indicated the delivery route using the Shandong Railway. The names in red are the post offices of the following usage examples.

民国9年中国郵便結束図
Postal Service Network in China, in 1920

〇印は、山東鉄道沿線の中国主要局
O: Principal Post Offices in Shandong Railway Zone

fig.2-2-1

三等局の郵便印

3rd Post Office Postmarks

fig.2-2.2 は民国 4 年 3 月 5 日に中国膠州局で受付けられた諸城宛の官製葉書です。高密局で中継され、3 月 6 日の中継印が押印されています。

「取極め」締結前で山東鉄道を利用できず、陸路を高密局経由で逓送されました。fig.2-2.3 の通りです。

郵便料金は国内宛葉書 1 分により 五色旗図 I 版 1 分が使われました。

ここで、郵便印の表示について説明します。fig.2-2.4 に 1911 年（辛亥）2 月 24 日付大清郵政通達 1 号における郵便局の区分による郵便印表示の通達内容を示します。この表示は中華郵政にも引き継がれました。但し、fig.2-2.5 の通り、中華郵政の郵便局区分は大清郵政から変更されました。

fig.2-2.2 の葉書に押印されている郵便印を fig.2-2.6 に示しました。中華郵政の区分により膠州局、高密局は共に三等局であったことがわかります。

fig.2-2.2 is a postcard to Zhucheng, accepted at the Chinese Jiaozhou PO on March 5, Minguo 4 (1915). It transited the Gaomi PO, with a transit mark dated March 6.

It was before the conclusion of the "Agreement," so they couldn't use the Shandong Railway and delivered mails by land transiting the Gaomi PO. fig.2-2.3 shows the route.

The postage was 1 fen for domestic postcard and it bears a 1-fen stamp of the 1st plate of five color flag.

Here I show the description on postmark. fig.2-2.4 shows the detail of notice about postmark of each post office category according to the Chinese Imperial Post Notice No.1 dated February 24, 1911 (Xinhai). The Republic of China Postal Service took over this description style. However, as you see in fig.2-2.5, the post office categories of the Chinese Imperial Post were changed in the Republic of China Postal Service.

fig.2-2.6 shows the postmark on the postcard in fig.2-2.2. In the category of the Republic of China Postal Service, Both the Jiaozhou PO and the Gaomi PO were the 3rd post offices.

fig.2-2-2

| '15.3.5 Chinese Jiaozhou PO | → 26km | '15.3.6 Chinese Gaomi PO | → 50km | Zhucheng |

fig.2-2-3

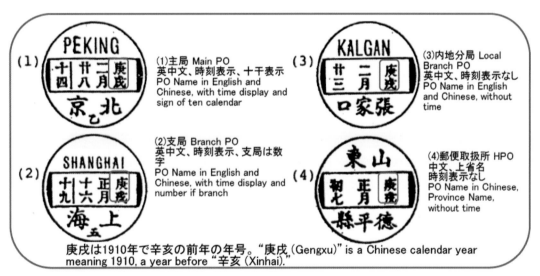

(1) 主局 Main PO
英中文、時刻表示、十干表示
PO Name in English and Chinese, with time display and sign of ten calendar

(2) 支局 Branch PO
英中文、時刻表示、支局は数字
PO Name in English and Chinese, with time display and number if branch

(3) 内地分局 Local Branch PO
英中文、時刻表示なし
PO Name in English and Chinese, without time

(4) 郵便取扱所 HPO
中文、上省名
時刻表示なし
PO Name in Chinese, Province Name, without time

庚戌は1910年で辛亥の前年の年号。"庚戌 (Gengxu)" is a Chinese calendar year meaning 1910, a year before "辛亥 (Xinhai)."

fig.2-2-4

大清郵政区分 Chinese Imperial Post Category	中華郵政区分 Republic of China Postal Service Category
(1) 主 局 Main PO	管理局:済南、一等局:青島、煙台、威海衛 Administrative PO: Jinan; 1st Class PO: Qingdao, Yantai and Weihaiwei
(2) 支 局 Branch PO	支局(数字)、二等局 Branch (number), 2nd Class PO
(3) 内地分局 Local Branch PO	三等局 3rd Class PO
(4) 郵便取扱所 Mail Handling PO (HPO)	郵便取扱所 Mail Handling PO

fig.2-2-5

丸二箱型印 Box Trisected Cds (Ø27mm)
使用期間 used between
'11.11.8 (高密/Gaomi)～ '23.7.3 (膠州/Jiaozhou)

膠州(三等局)
民国4.3.5.
Jiaozhou (3rd Class PO)
'15.3.5

高密(三等局)
民国4.3.6.
Gaomi (3rd Class PO)
'15.3.6

fig.2-2-6

fig.2-2.7 に民国 7 年 6 月 3 日に二十里堡郵便取扱所で受付けられ、済南、上海経由のニューヨーク宛の封書を示しました。郵便料金は国外宛封書により 10 分で 倫敦版帆船票 5 分が 2 枚貼付されました。

使用された郵便印を fig.2-2.8 に再現しました。郵便取扱所とは、郵便物も取扱う商店などを言います。従って、官が設置した郵便所ではありません。今で言うところのコンビニのようなところです。再現した郵便印にも中国語で「代辨所」と表示されています。官の代わりに郵便を扱うとの意味です。

fig.2-2.8 の郵便印は、石碑のような形状から碑型印と呼ばれています。年月日は手書きで記入されました。

日本と「取極め」締結後で山東鉄道を利用して済南に送られ、既に開通していた津浦鉄路、滬寧鉄路を利用して上海に逓送されました。

この封書の裏面に済南と上海の中継印が押印されています。両局とも丸一型の欧文日付印です。一見すると外国宛には丸一型印が使われたのかなとも思われるかもしれませんが、明確なルールがなかったようで、国内宛にも使用されています。

fig.2-2.10 に逓送路を示しました。東洋汽船は早くから太平洋航路の定期船を運航しています。サンフランシスコと横浜間に 2 回 / 月の間隔で運行し、ホノルルを経由しています。

fig.2-2.7 shows a letter to New York, accepted at the Er-shi-li-pu mail handling office on June 3, Minguo 7 (1918) and transited Jinan and Shanghai. The postage was 10 fen for international letter and it bears two 5-fen stamps of London print Junk.

The used postmark was reproduced in fig.2-2.8. A mail handling office was a shop etc. which handled mails in addition to its main job. It was not a post office established by the government. It was like a convenience store now. The reproduced postmark had a description "representative office" in Chinese. It meant handling mails for the government.

The postmark in fig.2-2.8 was called stone monument type postmark by its shape. The date was written by hand.

This letter was handled after the conclusion of the "Agreement" with Japan, so it was delivered by the Shandong Railway to Jinan, then to Shanghai by the Jinpu Railway and Huning Railway, which were already opened.

This envelope had transit marks of Jinan and Shanghai on the back. Both transit marks were bisected Roman Letter cancellations. At first sight, a bisected postmark was used for international mail, but it seemed that there was not a concrete rule and it was also used for domestic mails.

fig-2-2.10 shows the delivery route. Toyo Kisen had operated the regular service of the Pacific route from early on. They had two service monthly between San Francisco and Yokohama via Honolulu.

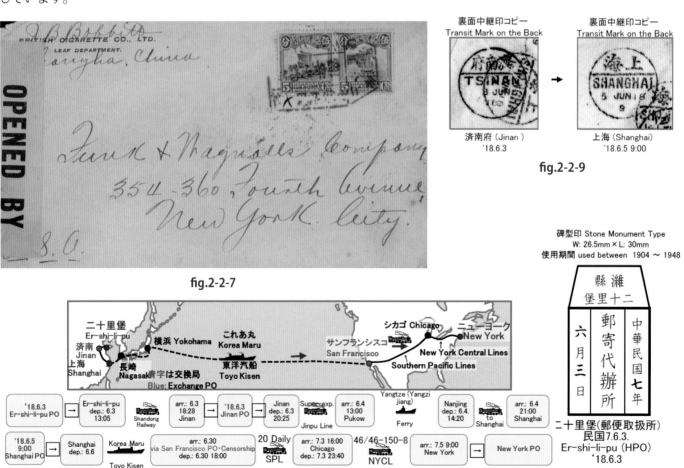

fig.2-2-7

裏面中継印コピー
Transit Mark on the Back　　　　　裏面中継印コピー
Transit Mark on the Back

済南府（Jinan）
'18.6.3　　　　　上海（Shanghai）
'18.6.5 9:00

fig.2-2-9

碑型印 Stone Monument Type
W: 26.5mm × L: 30mm
使用期間 used between 1904 ～ 1948

二十里堡（郵便取扱所）
民国7.6.3.
Er-shi-li-pu (HPO)
'18.6.3

fig.2-2-8

fig.2-2-10

済南局受付の上海宛の使用例を紹介します。fig.2-2.11 は民国 8 年 7 月 11 日に済南局で受付けられた上海宛の封書です。上海局に 7 月 13 日に届いています。郵便料金は国内宛封書 2 倍重量 6 分と書留 5 分の 11 分です。北京老版帆船票 1 分 2 枚と 3 分 3 枚が使われました。fig.2-2.12 に宛先面の縮小コピーを示しました。書留を示す印があります。

fig.2-2.13 に市内・国内・国外・青島宛の郵便料金表を示しました。日本軍撤退直前の 1922 年 11 月 1 日に郵便料金が値上げされました。ところが、国内宛が日本宛より高くなり、翌年 1 月 1 日に値下げされました。僅か 2 ヶ月間の値上げで、この間の中国内宛の使用は貴重なものとなっています。

I show here a usage example to Shanghai from the Jinan PO. fig.2-2.11 is a letter sent to Shanghai, accepted at the Jinan PO on July 11, Minguo 8 (1919). It arrived at the Shanghai PO on July 13. The postage was 11 fen consisting of 6 fen for double weight of domestic letter and 5 fen for registration fee. It bears two 1-fen and three 3-fen stamps of Beijing print old junk. fig.2-2.12 shows the reduced cover face with addressee. It had a mark of registered mail.

fig.2-2.13 shows the postage for a local and domestic mail and a mail to Qingdao. The postage was raised just before the evacuation of the Japanese troops, on November 1, 1922. However, the postage for domestic mail became higher than to Japan, and it was reduced on January 1 of the next year. The higher postage continued only two months and a usage of domestic mail in this period is so valuable.

fig.2-2-11

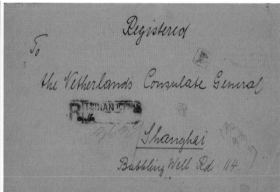

fig.2-2-12

宛先 Destination	市内 City Drop		国内 Internal		書留 Registration
基準 Standard Size	葉書 Postcard	封書 Letter 20g 毎 per 20g	葉書 Postcard	封書 Letter 20g 毎 per 20g	1897.11.24
1912.4.1	1分 1fen	1分 1fen	1分 1fen	3分 3fen	5分 5fen
1917.7.1	〃 as above	〃 as above	1½分 1½fen	〃 as above	〃 as above
1918.11.1	〃 as above	〃 as above	〃 as above	〃 as above	〃 as above
1922.11.1	1分 1fen	1分 1fen	2分 2fen	4分 4fen	7分 7fen
1923.1.1	〃 as above	〃 as above	1½分 1½fen	3分 3fen	5分 5fen

宛先 Destination	国外 International			青島宛 Qingdao		書留 Registration
基準 Standard Size	葉書 Postcard	封書 Letter 20g	20g 毎 per 20g	葉書 Postcard	封書 Letter 20g 毎 per 20g	1897.11.24
1912.4.1	4分 4fen	10分 10fen	1分 1fen	1分 1fen	4分 4fen	5分 5fen
1917.7.1	〃 as above	〃 as above	〃 as above	〃 as above	〃 as above	〃 as above
1918.11.1	〃 as above	〃 as above	〃 as above	1½分 1½fen	1.5g 毎 3分 3fen/15g	〃 as above
1922.11.1	1分 1fen	15分 15fen	8分 8fen	2分 2fen	4分 4fen	7分 7fen

fig.2-2-13 中国市内・国内宛郵便料金 / 国外・青島宛郵便料金 Internal tariff and outbound & Qingdao tariff

済南局の郵便印は丸二箱型印が使われています。fig.2-2.14に再現した郵便印を示しました。丸一型印から徐々に丸二箱型印に移行された時期の使用例です。fig.2-2.15に逓送路を示しました。済南から浦口までは津浦線、揚子江の連絡船で南京に渡り、そこから南海線（旧名称：滬寧鉄路）で上海に運ばれました。それぞれ特急列車で運ばれました。

中国青島局は日独戦で一時閉鎖されていましたが、民国7年10月に「山東日支通信連絡細項取極め」が調印されると11月に再開局されました。

fig.2-2.16に上海近傍の三林塘局で受付けられた青島宛の官製葉書を示します。fig.2-2.18に示した通り、海南線、津浦線、山東鉄道を利用し中国青島局に逓送されました。「取極め」により独自の逓送が可能となり、日本局の郵便印が押印されていないことから、日本局に交換されることなく配達されましたと考えられます。

郵便料金は青島宛の葉書1½分により 帆船図葉書国内用第2次1½分が使われました。中国青島局の使用例が少なく、3通の使用を確認していますが全て丸一型の郵便が使われています。fig.2-2.9に示した済南と同じ時刻表示のない郵便印です。日本租借時代の青島局の丸二箱型印は本品のみの確認です。

fig.2-2.17に再現した郵便印を示しました。三林塘局は時刻表示のない三等局の郵便印が使われました。fig.2-2.6の膠州、高密と同様です。青島局は時刻表示のある一等局の郵便印が使われています。

ドイツ租借開始時期は青島は郵便局がない、田舎の漁村でした。ドイツ、日本の租借時代を経て、山東省では済南に次ぐ都市になりました。

At the Jinan PO, box tri-sected cds was used. fig.2-2.14 shows a reproduced postmark. It was a usage example when the postmark was changing gradually from bisected type to box tri-sected cds. fig.2-2.15 shows the delivery route. It was delivered from Jinan to Pukow by the Jinpu Line, to Nanjing by a ferry, and to Shanghai by the Nanjing-Shanghai Line (former Huning Railway). The Super-express trains delivered it on each line.

The Chinese Qingdao PO was temporally closed because of the Japanese-German War, but reopened in November according to the "Agreement in Detail of Correspondence between China and Japan in the Shandong Province" which was concluded in October, Minguo 7 (1918).

fig.2-2.16 is a postal card to Qingdao, accepted at the Sanlin Tang PO near Shanghai. As you see in fig.2-2.18, it was delivered to the Chinese Qingdao PO by the Nanjing-Shanghai Line, the Jinpu Line and the Shandong Railway. The "Agreement" made it possible to for China to send mailslike this and this mail didn't have Japanese postmarks. So, it is supposed to have been delivered without exchange with Japanese post offices.

The postage was 1 ½ for postal card to Qingdao, and a 1 ½-fen of the 2nd junk domestic postal card was used. As for Chinese Qingdao PO, there are a small number of examples, and I have found only three items all of which were with a bisected postmark. And they were without time like that of the Jinan PO one shown in fig.2-2.9. By now, this is the only one example known of box tri-sected cds at the Qingdao PO in the Japan leased period.

fig.2-2.17 shows reproduced postmarks. The Sanlin Tang PO used the 3rd class PO mark without time. It was the same style of the Jiaozhou and Gaomi POs in fig.2-2.6. The Qingdao PO used the 1st class PO mark with time.

In the beginning of German occupation period, Qingdao was just a rural fishing village, which didn't have a post office. After the German and Japan leased period, it became the second biggest city next to Jinan in Shandong Province.

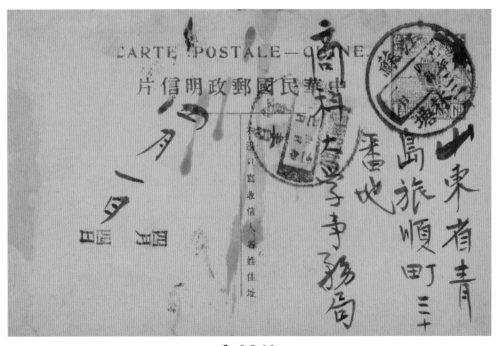

fig.2-2-16

但し、日本租借時代には青島には日本の郵便局がありましたので、業務を示す"十干"表示がないことから集配のみの業務に制限されていたと考えられます。

中国青島局は日本青島局の近傍の山東町にあり、窓口では受付をしていたと推察される使用例を確認しています。

In the Japan leased period, Qingdao had Japanese post offices. However, this postmark didn't have one of "ten calendar signs", which was a mark to identify the business counter of post office, so, this post office was supposed to have been limited to collect mails.

The Chinese Qingdao PO was established in Shandong town near Japanese Qingdao PO, and I have found a usage example which suggests they accepted mails at the counter.

丸二箱型印 Box Trisected Cds (Ø28mm)
使用期間 (済南 in Jinan)
used between '14.5.25 ~ '27.9.30

済南府
民国8.7.11. 18時
Jinan
'19.7.11 18:00

fig.2-2-14

丸二箱型印 Box Trisected Cds
(Ø30mm)

丸二箱型印 Box Trisected Cds
(Ø30mm)
開局期間(青島 Qingdao)
service between '14.1.1 ~ x.x, and;
(Closed temporarily because of
Japanese-German War)
'18.11.x, 1918 ~ '22.12.9

三林塘(三等局)
民国11.4.1.
時刻表示なし
Sanlin Tang (3rd Class PO)
'22.4.11, without time

青 島(一等局)
民国11.4.3.
時刻表示あり
Qingdao (1st Class PO)
'22.4.3, with time display

fig.2-2-17

済南→上海 逓送路
Route from Jinan to Shanghai

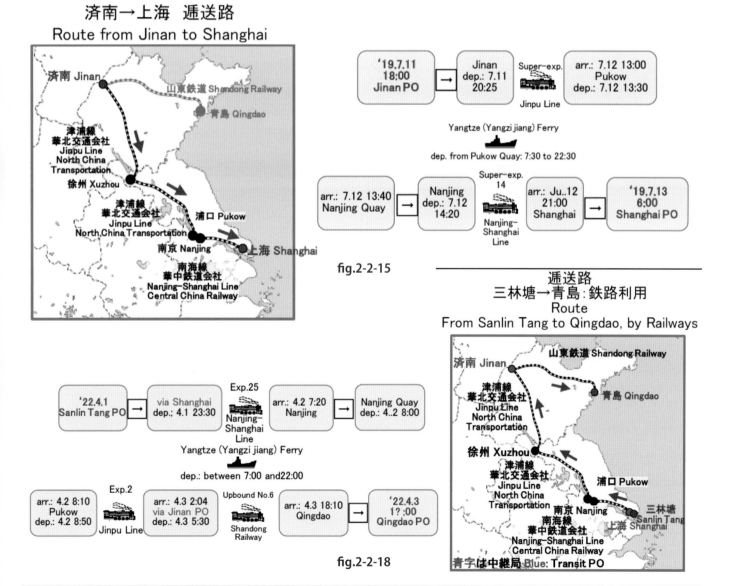

'22.4.1 Sanlin Tang PO → via Shanghai dep.; 4.1 23:30 — Exp.25 Nanjing-Shanghai Line — arr.: 4.2 7:20 Nanjing → Nanjing Quay dep.: 4..2 8:00

Yangtze (Yangzi jiang) Ferry
dep.: between 7:00 and 22:00

arr.: 4.2 8:10 Pukow dep.: 4.2 8:50 — Exp.2 Jinpu Line — arr.: 4.3 2:04 via Jinan PO dep.: 4.3 5:30 — Upbound No.6 Shandong Railway — arr.: 4.3 18:10 Qingdao → '22.4.3 1?:00 Qingdao PO

fig.2-2-18

'19.7.11 18:00 Jinan PO → Jinan dep.: 7.11 20:25 — Super-exp. Jinpu Line — arr.: 7.12 13:00 Pukow dep.: 7.12 13:30

Yangtze (Yangzi jiang) Ferry
dep. from Pukow Quay: 7:30 to 22:30

arr.: 7.12 13:40 Nanjing Quay → Nanjing dep.: 7.12 14:20 — Super-exp. 14 Nanjing-Shanghai Line — arr.: Ju..12 21:00 Shanghai → '19.7.13 6:00 Shanghai PO

fig.2-2-15

逓送路
三林塘→青島:鉄路利用
Route
From Sanlin Tang to Qingdao, by Railways

青字は中継局 Blue: Transit PO

以上で日本租借時代の説明を終わりますが、ここで租借についてドイツと日本の政策の違いについて述べたいと思います。産業革命で出遅れたドイツは遅れを取り戻そうと必死でした。産業革命後の欧州の鉄の生産量は第一位がイギリス、第二位がフランス、ドイツは第三位でした。そこで、製鉄するための石炭と製品の販売路として中国、特に、山東省に目を付けていたのです。

絶妙のタイミングでドイツの牧師が撲殺されるという事件が起こり青島に上陸、中国と戦闘らしきものがないままに租借条約を締結しました。その後、青島を近代都市にして、大型船が入港できるように青島湾を構築しました。上海をしのぐ貿易港にする狙いでした。ここから済南に鉄道を建設すれば中国の首都北京には上海より早く行けます。

全て予定通りに進みました。但し、ドイツは租借地と租借地外の地域を明確に分けて考えています。切手、郵便印で知ることができます。また、山東鉄道沿線局も濰縣と済南以外は全て閉局しました。

一方の日本は、ドイツが構築した郵便インフラを日独戦争で一瞬に手に入れました。そして、租借地外の郵便ネットワークを秘密局まで開局して充実させています。専管居留地を目指したわけです。

ドイツの郵便印は初めての郵便印は海軍野戦局、その後、地名が入り、時刻表示になります。日本も同様でした。日本はドイツと戦闘を行いました。中国と戦闘を行ったわけではありません。戦闘が終われば引き上げるべきところを居座りました。軍隊が駐留するので軍事局という名称を考えました。しかし、ドイツも日本も短期間の租借に終わり、目指していたことを実現できずに終焉しました。

That is the story of the Japan leased period. Now I tell about the difference of Germany and Japan about policy of leasehold. Germany, which got behind in Industrial Revolution, was frantic to keep up with other countries. After the Industrial Revolution, the country of the biggest volume of iron production in Europe was GB, the second was France and the third was Germany. So, Germany focused China, especially Shandong Province to mine coal for manufacturing iron and to sell the products.

When a German pastor was clubbed to death at a miraculous moment, the German troops landed on Qingdao and without any hard battle, concluded a leasehold agreement with China. After then, they made Qingdao a modern city and constructed the Qingdao Bay to make the vessels possible to land. They intended to make it a trade port bigger than Shanghai. If the railway was constructed there to Jinan, it would be possible to go to the capital city Beijing faster than from Shanghai.

Everything was going on as they had imagined. Germany clearly treated inside or outside of the leased areas differently. You can see which an area was included by checking its stamp and postmark. They also closed all the post offices in the Shandong Railway zone but Weixian and Jinan.

In Japanese case, they got at once the postal infrastructure constructed by Germany just by the Japanese-German War. And they developed the postal network out of the leased territories even establishing secret post offices. They intended to make their exclusive settlement.

The first German postmark was a naval field one, then, changed to those with location name and time. Japanese postmarks were also like that. Japan battled against Germany. They didn't battle against China. So, they should have evacuated China after the war but they stayed there. Since the military stayed, they created a name of "military PO." However, neither Germany or Japan kept their leasehold for a long time and realized what they had intended.

The Author： FUKUDA Shinzo

Collecting：Postal History of Qingdao & the Shandong Railway Zone

Awards:

Postmarks of Qingdao & the Shandong Railway Zone

　　LG + Grand Prix (JAPEX 2018, national exhibition)

Memberships:

　STAMPEDIA member since 2019

　Society for Promoting Philately member since 2019

Stampedia Philatelic Journal Index 2011-2020

How to order the back issues

This table shows titles and author-names of all the articles contributed to Stampedia Philatelic Journal.
You can find which edition contains the article in the right end number of edition = year.
The price of each issue including shipping cost is 20 EUR or 26 USD in total.
Please Email to **order@stampedia.net** We accept PayPal.

これまでの掲載記事全ての索引です。各記事の掲載号は、右端の年号数字で示されています。
バックナンバーは Amazon.co.jp にて、販売しているほか、下記店舗等で一部の号をお取り扱いいただいております。
切手の博物館ミュージアムショップ、ジャパンスタンプ商会、ユキオスタンプ、川口スタンプ社

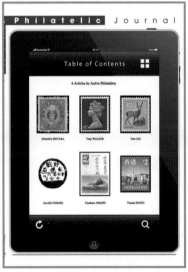

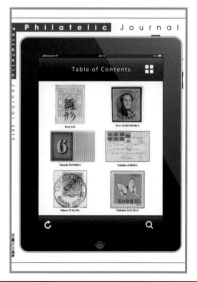

アジア＞伝統郵趣　Asia ＞ Traditional Philately		
初期北朝鮮切手に関する研究と収集 Study and Collection of Early North Korean Stamps	木戸裕介 Yusuke Kido	2015
インドのファーストイシュー（1852-1854） A Study of the First Issues of India（1852-1854）	プラグワ・コタリ Pragya KOTHARI	2016
インドステーツ・ハイデラバード Hyderabad	佐藤浩一 Koichi Sato	2017
アジア＞郵便史　Asia ＞ Postal History		
中国の開国　1745 ～ 1897 The Opening of China from 1745 to 1897	大場光博 Mitsuhiro Ohba	2011
1873 年～ 1875 年のダフラ封鎖・ダフラ野戦軍の軍事郵便 ～ 英領インド北東辺境地域における軍事郵便の例 ～ Military Mails of Dafla Blockade and Dafla Field Force in 1873-1875 Examples of Military Mails in North East Frontier Tracts of British India	小岩明彦 Akihiko KOIWA	2016
青島局と山東鉄道沿線局の郵便史 ドイツ租借時代：1898-1914 年 Postal History of Qingdao and the Shandong Railway Zone German Occupation Period: 1898 to 1914	福田真三 Shinzo FUKUDA	2019
青島局と山東鉄道沿線局の郵便史 日本租借時代：1914 年～ 1922 年 Postal History of Qingdao and the Shandong Railway Zone, Japan Leased Period: 1914 to 1922	福田真三 Shinzo FUKUDA	2020
ヨーロッパ＞伝統郵趣　Europe ＞ Traditional Philately		
エストニア花図案切手収集の楽しさと難しさ Appeal and Difficulty of collecting the Estonian Flower Design Stamps	伊藤昭彦 Akihiko Ito	2011
オーストリアのファースト・イシュー / トラディショナル収集 First Issue and Traditional Collection in Austria	斎藤環 Tamaki Saito	2011
バーゼルの鳩，複数の印刷版の存在をつきとめた研究 Basel Dove, First Time Proof of the Existence of Two Different Plates	カール ルイス Karl LOUIS	2013
戦後の切手で国際展を戦えるか？ CAN MODERN PHILATELY COMPETE SUCCESSFULLY AT INTERNATIONAL LEVEL?	アンソニー ウォーカー Tony WALKER	2014
バイエルン数字図案シリーズ 1849-1862 ～プレーティング及び新版の可能性について～ Bavarian Numeral Issue 1849-1862- Plating and Possibility of Existence of New Plates –	福田達也 Tatsuya FUKUDA	2016
エメラルド コレクション The Emerald Collection	デイビッド フェルドマン David FELDMAN	2018
メーメル地区の歴史 The History of the Memel Region	トビアス ヒュルマンス Tobias HUYLMANS	2018
ベルギー メダリオン切手の「円形」変種 The Belgian "Medallion" issue – October 1849 till June 30 1866, The "Circle" varieties	イヴ・フェルトメン Yves VERTOMMEN	2019
チューリッヒ 6 Rp. リコンストラクションへの挑戦 A Challenge to Reconstruct Zürich 6 Rappen	吉田 敬 Takashi YOSHIDA	2019
ヨーロッパ＞郵便史　Europe ＞ Postal History		
超インフレカバー収集の楽しさ Appeal of collecting Hyperinflation Covers	伊藤文久 Fumihisa Ito	2012
フランス革命戦争・ナポレオン戦争期の英仏間郵便史 FRANCE – UNITED KINGDOM: A TRAVEL ACROSS THE MAIL COMMUNICATION 1793-1815	ルカ・ラヴァニーノ Luca Lavagnino FRPSL	2015
1849-1852 ジュネーブにおける貨幣のスイスフランへの移行 1849-1852 How the "French Currency" of Geneva became the Swiss Franc	ジャン・ヴォルツ Jean Voruz	2015
香港からアウシュヴィッツへ Hong Kong to Auschwitz	内藤 陽介 Yosuke NAITO	2018
アメリカ＞伝統郵趣　America ＞ Traditional Philately		
ハワイ初期切手 Hawaii early stamps	山崎文雄 Fumio Yamazaki	2011
ハワイ数字切手におけるバラエティーの研究 Study of Varieties on Hawaiian Numerals	山崎 文雄 Fumio YAMAZAKI	2018
アルゼンチンのファーストイシュー切手について The First Issue Stamps of Argentina	佐藤浩一 Koichi Sato	2011
「牛の目」の初日使用例 First Day Use of BULL'S EYES	正田幸弘 Yukihiro SHODA	2016
ブラジル 1844 年シリーズ高額切手の使用例 Usages of High Face Value Stamps of Brazil 1844 issue	正田幸弘 Yukihiro SHODA	2017
海外宛「牛の目」シリーズ使用例 "Bull's Eye" Issue, Overseas Usages	正田 幸弘 Yukihiro SHODA	2018

書　名：スタンペディア　フィラテリック　ジャーナル　２０２０
巻　数：第 10 巻（2020 年版）
発行日：2020 年 11 月 10 日
価　格：１，１００円（消費税込）
発行部数：600 部
発行者：無料世界切手カタログ・スタンペディア株式会社
発行人：吉田敬
編集部：北川朋美、菊地恵実

Name of the magazine:	Stampedia Philatelic Journal 2020
Volume:	10th (2020 edition)
Date of issue:	November 10th 2020
Price:	1,100 Yen
Number of printing:	600 copies
Publisher:	Stampedia, inc. YOSHIDA Takashi
Editor:	KITAGAWA Tomomi, KIKUCHI Emi